The
Hermetic Code
in DNA

The Hermetic Code *in* DNA

The Sacred Principles in the Ordering of the Universe

Michael Hayes

Inner Traditions
Rochester, Vermont

Inner Traditions
One Park Street
Rochester, Vermont 05767
www.InnerTraditions.com

Originally published in the United Kingdom in 2004 by Black Spring Press under the title *High Priests, Quantum Genes*

Library of Congress Cataloging-in-Publication Data
Hayes, Michael, 1949–
  The hermetic code in DNA : the sacred principles in the ordering of the universe / Michael Hayes.
     p. cm.
  "Originally published in the United Kingdom in 2004 by Black Spring Press under the title *High Priests, Quantum Genes*."
  Includes bibliographical references and index.
  Summary: "An examination of the precise code that connects ancient spirituality with modern science"—Provided by publisher.
  ISBN: 978-1-59477-218-4
  1. DNA. 2. Spirituality. I. Hayes, Michael, 1949– High priests, quantum genes. II. Title.
  QP624.H39 2008
  572.8'6—dc22

                                                                    2008004438

Printed and bound in the United States by Lake Book Manufacturing

10 9 8 7 6 5 4 3 2 1

Text design and layout by Jon Desautels
This book was typeset in Garamond Premier Pro with Bauer Bodoni as a display typeface

To send correspondence to the author of this book, mail a first-class letter to the author c/o Inner Traditions • Bear & Company, One Park Street, Rochester, VT 05767, and we will forward the communication.

Contents

Foreword

I suspect that the name of Michael Hayes is going to be remembered together with those of Stephen Hawking and Watson and Crick as a thinker who has made a revolutionary contribution to our vision of modern science.

Some time in 1995 I received a copy of a book called *The Infinite Harmony,* and subtitled *Musical Structures in Science and Theology,* published by the respectable firm of Weidenfeld & Nicolson. Since I was overworked, trying to complete a book to a deadline, it took some time before I got around to reading it. My book was about ancient Egypt and was called *From Atlantis to the Sphinx;* its starting point was the theory of John Anthony West that the Sphinx may be thousands of years older than anyone had supposed. And the amount of reading required was enormous.

One evening I was relaxing with a glass of wine when I noticed *The Infinite Harmony* in a pile of books beside my chair. I picked it up idly, glanced down the table of contents, and saw that the second chapter is devoted to ancient Egypt. Naturally, I turned to it immediately, and was soon reading with excitement and absorption. I quickly learned something I had not come across before: that in the antechamber to the King's Chamber in the Great Pyramid, there is a square granite relief whose area is exactly equal to the area of a circle whose diameter happens to be precisely the same length as the antechamber floor. What is more, when this length is multiplied by *pi,* the result is precisely the length of the solar year: 365.2412 pyramid inches.

I was fascinated. It had long been clear to me that the ancients attached some mystical significance to numbers and that the sophistication of their knowledge was often greater than ours. Hayes reinforced my feeling that we are dealing with a very ancient knowledge system whose secret has been lost.

I was so excited that I looked around to see if I could locate the letter that had accompanied the book. It had vanished. The inscription in the book showed that it had been lying around my sitting room for months. And my wife had made a note of the sender's address, which was in Moseley, Birmingham. I rang Directory Enquiries and asked them if they had a telephone number for Michael Hayes; they had. And although it was now after ten in the evening, I rang him. A girl answered the phone, and went off to get her father. A few moments later, I was speaking to Michael Hayes, apologizing for keeping him waiting so long for a reply, and telling him that I found his book enormously exciting.

I asked him some questions about himself, and about how he had become interested in the subject. He told me that it had started in his hippie days, when he was living in Mashad, in Iran, and was in the great mosque of the Imam Reza, impressed by the sheer number of worshippers, and by their devoutness. It was obvious that to them, religion was a living reality, just as it had been to the thousands of worshippers who had brought stones for the building of Chartres cathedral in the twelfth century. And during his travels in Iran, India, Pakistan, and Afghanistan, Michael Hayes had felt exactly the same thing—that their religions had a living source. He experienced an overpowering sense of being on the brink of learning some enormous secret.

Back in England, he had decided that it was time he learned something about the genetic code, and the mysterious letters DNA. He enrolled at a course at Leicester University. And there he took an important step closer to the secret. It proved to be numerical.

The spiral-shaped DNA molecule involves four chemical bases

called adenine, guanine, cytosine, and thymine. And these four can combine together in sixty-four different ways to form triplet units called RNA codons.

The number 64 struck a chord. Then he remembered what it was: that the Chinese "Book of Changes," the I Ching, has sixty-four "hexagrams," each made up of two different lines. Any reader who has ever tried throwing down three coins to consult the I Ching will recall that a preponderance of tails result in a broken line,

----

while three heads form an unbroken line:

————————————

The first symbolizes the Chinese concept of "yin," the feminine, the yielding, while the second is "yang," the forceful and masculine. The coins are thrown six times, and the six lines are laid on one another in a kind of six-decker sandwich.

Those who use Richard Wilhelm's translation, with the introduction by Jung, will recall that the next step is to turn to the chart at the back of the book, which contains sixty-four numbers in a grid of squares, whose sides are eight units long. You then look up your "top" trigram along the horizontal edge, and your "lower" trigram along the vertical edge, and the square where the two trigrams meet is the number of the hexagram you are looking for.

In the early stage of his quest, Mike Hayes (as he prefers to be known) had studied the I Ching, and wondered idly why the number of hexagrams is eight times eight, not seven times seven or nine times nine. And now, with the coincidence of the DNA code and the hexagrams of the I Ching, he found himself wondering if this number 64 is some basic code of life.

When he learned that there were eight trigrams hidden in DNA, he began to feel that this was more than an odd coincidence. . . .

All this Mike sketched out for me during that phone conversation. And when it was over, I had decided that reading the whole book was a major priority.

What I learned in *The Infinite Harmony* was that this coincidence was just the beginning of a whole series of related discoveries. For example, the number 22 plays a basic part in the DNA code. Proteins are formed by twenty amino acids, but with two codons forming start and stop signals, making twenty-two in all. And 22 also plays an important part in music, being the number of notes in three octaves on the piano. The followers of the Greek philosopher Pythagoras regarded 22 as a sacred number, and also 3.

Previously, studying the Russian mage Gurdjieff, Mike had also been introduced to something called the law of three. Positive and negative, good and evil, light and darkness, merely counterbalance one another, but a third force is necessary to combine them—just as the two sides of a zipper are made to interlock with the fastener in the middle, or two gases will only combine in the presence of a catalyst that is itself unaffected.

Studying the world's major religions, Mike was struck by how often the numbers 22, 3, and 7 occur. The number *pi,* the relation of a diameter of a circle to its circumference, is 22 divided by 7. So now he began to look in detail at the world's major religions—ancient Egyptian, Judaism, Zoroastrianism, Islam, Jainism, Buddhism, Confucianism, and Christianity. With increasing excitement, he realized that his numerical discoveries constituted a code that connected them all. The same code turned up in alchemy, which led him to label it the Hermetic Code, after Hermes Trismegistus, the Greek god who is the patron of alchemists, and whose best-known dictum is "As above, so below." And so *The Infinite Harmony* came to be written.

His chances of publishing such a strange and abstruse book seemed minimal, yet its importance was recognized by an editor at Weidenfeld & Nicolson, and it appeared in 1994. But there the marvelous wave of coincidence and synchronicity that had carried him so far seemed to run out of strength. The book was not widely recognized, and opened no further opportunities. And just as Mike was beginning to experience a sense of anticlimax, I rang up, and said I intended to write about it in *From Atlantis to the Sphinx.*

I did just that, and the book came out in 1996, and went into several editions—partly because the whole subject of ancient civilizations had become popular as a result of Graham Hancock's remarkable bestseller *Fingerprints of the Gods,* which argued that civilization may be thousands of years older than archaeologists believe.

By that time, I had met Mike Hayes. He had spent a part of his childhood in Penzance, in Cornwall, and accepted eagerly when I suggested that he should take a few days off writing his second book, and come and renew his acquaintance with Cornwall. We spent days driving around, talking endlessly, and he told me many things about himself and his development that I shall not repeat here, since they are in the remarkable and absorbing introduction to this book.

Mike proved to be a slightly built, fair-haired man who was in his mid-forties at the time I met him. And during the few days he spent in Cornwall (his wife, Ali, had to stay behind to look after their three daughters), I got the same odd feeling I had experienced while reading *The Infinite Harmony*: that here was one of those people that fate seems to throw down into the world to make some important discovery.

This has always seemed to me true of all scientists and inventors. One of my favorite television programs is Adam Hart Davis's *Local Heroes,* in which he cycles from place to place, and comes upon dozens— in fact hundreds—of remarkable men and women who have left something behind them, perhaps something as straightforward but essential as the lawnmower or hovercraft, perhaps some world-changing knowledge like relativity or quantum theory.

Mike Hayes, I soon came to feel, is one of these.

And why do I think he is so important? Because if the genetic code and Mike's Hermetic Code—these numbers that recur constantly throughout all world religions—are identical, then there is a fundamental connection between molecular biology and religion. And why is that important? Because ever since Gregor Mendel created genetics in the nineteenth century, it has been regarded as a science of the mechanism of evolution. Darwin suggested that evolution progresses through a mechanical process of the survival of the fittest, but he was not sure about the nature of the mechanism that creates species. Mendel's discoveries pointed to the genes as the answer.

But Darwinism plus Mendelism was even more mechanical than Darwinism alone. At least Darwin believed that his colleague Lamarck— and his grandfather Erasmus—might be partly correct in believing that the will of the individual influences evolutionary changes. But the neo-Darwinists who accepted Mendel's discoveries as the mechanism of evolution felt that it explained everything. Evolution was now a totally mechanical process—like the erosion of a landscape by geological forces—for the will of the individual cannot influence his genes. And the most influential of modern geneticists, like Richard Dawkins, are rigid materialists.

I personally have been attacking this view for the past half century, and have pointed out anomalies that cannot be explained in terms of mechanical evolution—for example, how a colony of little insects called the flattid bug can crawl onto a dead twig and then shape themselves into the likeness of a living flower—a flower that does not even exist in nature. This cannot be explained by "survival of the fittest." It seems to involve some "group mind" operating at an unconscious level.

Now, in showing the connection between the Hermetic Code and the genetic code, Mike Hayes has pointed to the fact that the essence of evolution can also be found in religion, and therefore in the realm of the evolution of consciousness.

I found his introductory remarks about the insights he obtained through LSD exciting partly because of his comment, "I clearly perceived that (everything solid) is composed, literally, of sparkling, vibrant 'particles' of light," a view that is of central importance to the argument of the book, and that echoes the vision of so many mystics.

Now, I had already come upon this notion in a book called *Essay on the Origin of Thought* (1974) by a remarkable young philosopher named Jurij Moskvitin. Lying one day in the sunlight with his eyes half closed, he became aware of a kind of moving mosaic pattern through his eyelashes. It seemed to be made of tiny light fragments, and as he slowly developed the ability to focus them, he recognized patterns like those in religious art, "art and ornamentation created by civilizations dominated by mystical initiation and experience." These forms, he finally decided, were made up of "dancing sparks," a little like the tiny lines in the work of the painter Signac. These sparks, which he decided looked a little like tadpoles, make up our whole visual field, on which we impose shapes. He compares it to the way that, in a Dutch painting, a wineglass examined closely proves to be merely a few strokes of yellow paint. Moskvitin is suggesting that the external world our eyes reveal to us is simply a limited version of a larger inner world. I was reminded of Moskvitin's thesis by Mike Hayes's theory of light—on which he expands greatly in this book.

His insights were also close to those of a remarkable anthropologist called Jeremy Narby, who studied among the Ashaninca Indians of Peru, and became convinced that their extraordinary knowledge of the medicinal properties of forest plants was obtained through a visionary process involving the drug ayahuasca.

For example, the drug curare, used on poison darts, is made from a combination of plants, and the first stage is to boil them for three days, while staying clear of the deadly vapors. The final product kills monkeys without poisoning their meat, and also causes them to relax their grip so they fall from the tree to the ground, instead of clinging to the tree in a death spasm.

But there are about eighty thousand species of forest plants. How did the Indians stumble on curare without poisoning themselves first, or wasting their lives in endless experiment?

The same questions arise with regard to ayahuasca. It is made up of two plants, one of which contains a hormone secreted in the human brain, a hallucinogen that is rendered harmless by a stomach enzyme. In order to prevent it being rendered harmless (and useless as a drug), it has to be mixed with a substance from a creeper. Then it induces visions.

How, Narby wondered, did the Indians discover anything so complex? Surely not by trial and error—trying millions of possible combinations. The shaman's answer was that they learned it from drugs, which "told" them the answer.

Narby learned a great deal from another anthropologist, Michael Harner, who had also experimented with drugs among the Indians. And Harner had declared that his visions emanated from giant reptile creatures "like DNA" that resided at the lowest depth of his brain.

It struck Narby that DNA looks like two intertwined serpents (as Mike Hayes also points out). The molecule also looks like a spiral ladder, and shamans the world over talk about ascending a ladder to higher realms of the spirit.

Narby himself tried ayahuasca, and reached the same conclusions as Harner. The drug introduced him to Harner's "serpents":

> Suddenly I found myself surrounded by two gigantic boa constrictures that seemed fifty feet long. I was terrified. . . . In the middle of these hazy thoughts, the snakes start talking to me without words. They explain that I am merely a human being. I feel my mind crack, and in the fissures, I see the bottomless arrogance of my presuppositions. It is profoundly true that I am just a human being, and, most of the time, I have the impression of understanding everything, whereas here I find myself in a more powerful reality that I do not understand at all and that, in my arrogance, I did not even suspect existed.

He began to feel that language itself was inadequate, and that words would no longer stick to images.

But after this alarming beginning, things began to improve as he realized that the Indians know their way around in this bizarre reality, and that the most apparently absurd things they had told him were true. And somehow, the Indians seemed to be obtaining their information directly from DNA, a concept that seems less odd when we remember Mike Hayes's discovery of the similarity between the genetic code and the I Ching.

Later in *The Cosmic Serpent,* Narby writes, "It seemed that no one had noticed the possible links between the 'myths' of 'primitive peoples' and molecular biology." And he goes on to make the important comment (in view of Mike Hayes's emphasis on music), "According to the shamans of the entire world, one establishes communication with the spirits via music."

Narby dares to ask, "Is there a goal to life? Do we exist for a reason? I believe so, and I think that the combination of shamanism and biology gives undisputed answers to these questions."

Obviously, Jeremy Narby and Mike Hayes have been pursuing parallel courses, and arrived at very similar conclusions.

A few words about the present book.

In many ways, it is easier to absorb than *The Infinite Harmony.* To begin with, Hayes discusses in his introduction the pertinent biographical facts that enable the reader to watch the discovery and unfolding of his ideas. This introduction says everything that is in *The Infinite Harmony,* and makes it all beautifully clear. He then plunges into the questions that are directly related to Graham Hancock's thesis in *Fingerprints of the Gods,* Robert Bauval's in *The Orion Mystery,* and my own in *From Atlantis to the Sphinx.* Even I, who have now devoted about ten years to these matters, was fascinated by his treatment of them. He also points out that there is evidence that Neanderthal man knew about the Hermetic Code seventy-five thousand years ago.

I shall not try to summarize the rest of the book except to say that it is remarkable for the confidence he shows in handling an immense range of subjects, from modern physics to the paranormal, from evolutionary biology to musical theory, from yoga to superconductivity. I was familiar with some of this material, but much of it was unknown to me, and the use he makes of it is strikingly his own.

The performance is often so dazzling, reminding a reader of a juggler who can keep ten balls in the air at the same time, that the reader might easily be misled into thinking that this is no more than a brilliant piece of eclectic exposition. But make no mistake: what Mike Hayes has discovered could be as important as the original discovery of DNA. Like Jurij Moskvitin and Jeremy Narby, he has created a new paradigm—that is, he is looking at our familiar universe from a new angle, and making us aware of magical possibilities.

COLIN WILSON

Colin Wilson is a prolific author and philosopher whose 1956 breakout work *The Outsider* helped popularize existentialism in Britain. Later, when existentialism fell out of fashion, he became a symbol of the British version of the beat generation as a member of the "Angry Young Men," in which he was the head of a small group of existentialist philosophers. Beyond his early political influence, his more than 108 titles convey his enormous literary scope—ranging from philosophy, crime, occult, literary criticism, and short fiction—and include *From Atlantis to the Sphinx, Atlantis and the Kingdom of the Neanderthals,* and his autobiography, *Dreaming to Some Purpose.* He is also coauthor, with Rand Flem-Ath, of *The Atlantis Blueprint.*

Acknowledgments

I would like to thank Colin Wilson for all the help and encouragement he has given me in the writing of this book, and for always finding time in his busy schedule to answer my calls. A kinder, wiser man I have yet to meet. I am duty bound also to thank posthumously three other wise men who have helped shape my world: George Gurdjieff, Pyotr Ouspensky, and Rodney Collin. Without their input, I should never have dreamt of such wonderful things.

A special thanks to Kay Hyman, whose invaluable editorial contribution has been generously provided simply for the love of it.

And lastly, but most of all, thanks to my wife, Ali, for reasons too numerous to mention.

A Note on Measurements

When taken from other sources, units of measure used in the book retain the measurement system used in the original text. So temperatures may be in Kelvin, Celsius, or Fahrenheit, and physical measurements may be metric or imperial, and so on.

Introduction

This book is the product of a personal journey of discovery, a trip that began when I was about seven or eight years old. This, significantly, was when I first chanced to think about that ultimate question in life: death. I remember feeling greatly disturbed that I was unable to comprehend this truly awesome prospect. What made matters worse was the fact that the adults around me were not only equally clueless in this respect; it seemed to me that they didn't even want to think about it. But then this was England in the mid-fifties, and the grown-ups had just survived a horrendous global war. For many of them the unspeakable facts of death must have been an all-too-prevalent and uncomfortable reality, so it is not surprising that I was usually given short shrift whenever I asked one of the available big people to show me the netherworld on a world map.

As it turned out, and for reasons I cannot explain, I have been drawn to ponder this question many times over many years. So, if nothing else, the subject has been a recurrent reminder to me of the transient, apparently futile nature of individual existence. But it has also, I think, been a primary factor in determining one of my major motivations in life—to try to understand the meaning and purpose of our being, to establish some kind of meaningful perspective from which to view our true position in the cosmic scheme of things. Basically, I simply want to know what is going on around me. Don't you?

1

So, what is this thing death, this future happening looming over the horizon of our lives like some conceptual black hole? Can the process be elucidated, defined in terms we can understand? My answer is a cautious yes, and I shall explain why in due course.

As for death's equally mystifying opposite, the counterbalance we loosely call life, this too cries out to be understood. Evolutionists think they have cracked it by charting the increasingly complex interactive development, over four thousand million years, of the RNA and DNA molecules—which is fine, as far as it goes—but where does the evolution of consciousness fit into the Darwinian picture? Indeed, can it fit? That is, is it possible to explain the thought processes of the modern hominid in terms of the current theory of evolution? Actually I don't personally know of any evolutionists out there who are aware of this fact, but the answer, once again, is yes. As I see it, the systematic, biomolecular process involved in the evolution of DNA is a perfect model of the working of the healthy human mind.

So what I currently have to offer is an ambitious, but serious, proposition, which is that life and death are in a certain and unique way entirely comprehensible.

As I said, what follows is the record of a personal journey, but this is also, by its very nature, an account of the entire evolutionary journey of the conscious hominid. What began for me in the fifties, with what might be called a chance thought, has been happening to thinkers for many thousands of years. So, in effect, I have merely tuned in to an already existing stream of ideas, a channel of intelligent information whose list of presenters and past contributors reads like a roll call of the immortals—scientists, philosophers, saints, mythmakers, saviors.

We will therefore have to go back in time to trace the origin of this "thought" of mine: back to ancient Egypt and Greece, and to China, India, Palestine, Arabia, and the Americas. One of the principal reasons for looking back is that most of these ancient cultures developed a religion, or a mythology, to explain the mystery of life and death.

9000,000,000

1,000,000

Indeed, this almost wholly preoccupied the earliest peoples. And, significantly, although different cultures over the millennia have expressed their ideas in apparently diverse ways and idioms, they all agree on one fundamental point: that there is an existence after death. As it happens, the originators of all the major belief systems also concurred on one other fundamental point in respect of life.

But first things first. My own personal account is the warp of this metaphysical design, so we must for the time being stick to the minor plot; the greater weft will be woven in chapter by chapter.

In the early sixties, I dutifully went to grammar school, obtained mediocre GCE passes, and subsequently took up a position selling advertising space for a local newspaper. Disillusionment soon crept in. A large workplace can be a quagmire of trivia and petty jealousies and, to avoid being sucked in, I became a corporate drifter, aimlessly career-hopping from one meaningless job to another.

Meanwhile my alter ego was heading off on a completely different trail. By the time of the late sixties, he was already blowing in the wind, unwittingly heading for a second memorable jolt. This happened when, quite by chance, I came across a certain psychedelic agent called purple haze.

Purple haze was the name given to a particularly pure batch of LSD that hit the streets of my town in the winter of 1968, one tiny tablet of which happened to come my way. It cost me thirty bob and about eight earth hours, but such was its impact upon me that it changed my whole life, for, quite suddenly, after this one, mind-blowing experience, I became absolutely convinced of the existence of other dimensions beyond my own tiny, subjective conceptual domain. This newly found awareness made life appear much more interesting. But more perplexing.

More trips inevitably followed, always, without exception, profoundly illuminating, producing in me such powerful waves of emotion that I felt I could very easily be swept out to some mystic sea and be gone forever. Whether these glimpses into other worlds were real or imaginary was a question I never bothered to ask, but my perceptions were so vivid and incomparably impressive that they made my molelike working life seem like a form of penal servitude.

It has been more than twenty years since I last took a trip, and I have no intention of taking another in the foreseeable future. Neither do I recommend the use of psychedelics to anyone. I am merely reporting here. My own "transgressions" were directed largely by circumstance. This thing—this drug—was new and radical, and virtually everyone in my peer group was experimenting with it. Obviously, in another time and another place, with different peers, I might have taken an alternative route to the present.

So, to get to the point, which is to explain why my psychedelic experiences should be of such importance to my story. It has all to do with the impressions I had then. To be sure, very little remained of the total experience after each of these illicit forays into inner space, but certain key impressions did remain indelibly imprinted on my mind.

The first was that everything solid or material—houses, trees, rocks, mountains, people—were all possessed with a kind of inner light of their own. That is, I clearly perceived that these things, or objects, were composed, literally, of sparkling, vibrant, "particles" of light.

It is entirely possible, of course, that this is not so—that "things" are not composed of light at all—and that the impression was simply a drug-induced false consciousness. However, when the hallucination, or whatever it is, appears to be infinitely more striking and meaningful than anything so-called reality can throw up, then I think I have good reason to pay heed to it—which, indeed, I have done ever since. And, in fact, although I was unaware of this at the time, I was later to discover that my impression was corroborated by two quite different and independent sources.

In the first instance, Einstein had already shown that light quanta, i.e., photons, were "particles." Second, the idea that matter is simply one particular form of light has been common currency among the holy men of the East for centuries.

The second major impression (or hallucination) had to do with time—or, rather, the absence of it. I could never explain it, not even to myself, but in these altered states time seemed to stand still. I remember that the word *eternal* came to mind more than once when I was attempting to describe this condition.

Interestingly, this particular notion—that there are "timeless" realms, or dimensions, of existence—is not at all unique. In fact, it is part and parcel of practically every major religion and mystical belief system known. If you think of familiar scriptural concepts like heaven, eternity, time without end, the realm of Him that liveth for ever and ever, and so on, all these so-called religious notions seem to suggest that legions of contemplatives have in the past had glimpses or feelings similar to mine. Further, as with the earlier impression that matter is made up of vibrating particles of light, this second idea of a "timeless" form of reality has also been quite clearly expressed in independent sources. For example, through the development of modern quantum theory, it has been discovered that, in the "world" of the subatomic particle, time as we know it (or as we think we know it) has no place: it is statistically meaningless.

For me, this idea of an "eternal" dimension of existence was especially appealing, because it seemed to hint at a possible way out of the time-laden quicksand in which we hapless mortals become immersed. That is, if there was any substance at all to my extratemporal experiences, then maybe we—you, me, everyone—need never truly die.

I have always been an avid reader, but over the years my taste changed with my circumstances. So, by the end of the sixties, works of fiction, classic and popular, were gradually replaced by books on science and what my elder brother Tony laughingly called "all that esoteric stuff." He was right, of course. The hippies were on the move, traveling in droves to the East, reading books by and about countless Indian holy men, Sufis, Western occultists, and Lobsang Rampa and Erich von Däniken to boot. I readily joined in the party, reading all kinds of spiritual and philosophical fare. Much of it I found pretty ineffectual: hearsay, vague allusions, apocryphal stories, parables, and outright guesswork—but all in all I was temporarily hooked, greatly impressed by the vast numbers of mostly sincere writers from all walks of life attempting to understand the nature of consciousness. Of all intellectual pursuits, the exploration of the human mind seemed to me to be the most worthwhile. If we could reach journey's end on this one, all other questions might fall neatly into place, side by side with their answers.

The trouble was, although a lot of the books circulating in the seventies contained many interesting ideas, after reading them I still had no idea what was really going on in people's heads. So many writers claimed to have all the answers, but when it came down to the nitty-gritty, everything seemed to end with a question mark. I had no inkling then that a major clue was in the offing, but I was soon to find a man who had some important answers. What is more, he wasn't entirely unique.

Usually with my wife, Ali, I made several trips to the East during the seventies. Often we would stop off and visit my brother, who at that time was living in Mashad in northeastern Iran. Tony, who never stopped traveling throughout the whole of his abbreviated life, had at this stage in his journey married and converted to Islam.

This was in the days of the pro-Western Shah, and Iran had a booming economy, affording plenty of opportunity to anyone with an entrepreneurial flair. And yet, despite all this, the people remained deeply religious, especially in Mashad, one of Iran's holy cities, home of the great mosque of the much-venerated Islamic saint Imam Reza, the fourteenth imam in a direct line of high initiates that began with the Prophet Muhammad himself.

During these visits, I was always struck by the intense fervor and passion of Muslim worshippers there. Their tears were obviously very real, and their emotions seemed to be charged with a vitality of a kind seldom encountered in Christianity. To these people, prayer was a genuine, wholehearted celebration, a loud, proud, public affirmation of their devotion to Allah and His Prophet.

I must admit that my interest in Islam, although it impressed me greatly, never passed beyond an observational level. What intrigued me most about it was the sheer emotional power that this metaphysical phenomenon had so effectively harnessed. There was a self-evident force at work here—not a force that could be empirically measured according to established scientific criteria, but a real source of power nonetheless, one so energetic, in fact, that it could somehow cause millions of people from different ethnic backgrounds all over the world to simultaneously move, speak, and act in concert. Perhaps the most curious thing about this remarkably well-coordinated mass movement of human beings is that it was all set in motion by one man.

When I try to picture Muhammad in my mind, I see a person of true genius, the light of Allah sparkling in his eyes, a clear vision of the future march of Islam stretching out before him. Here, quite evidently, was a man who knew exactly what he was doing, an individual who understood the workings of the human mind like few others. How else could he have created such a powerful living movement? Luck? Accident? I really don't think so. There is a weird kind of magic afoot here, and it comes to us today in the form of a tangible supernatural force—the mysterious power of Islam. Now, this force exists, it cannot be denied, and I am saying that the person who purposefully created it was—and indeed still is—without doubt a giant among men. Irrefutable evidence to support this view is provided daily, weekly, continuously for all to see, when millions of Muslims all over the world emulate their leader by taking time to align themselves with this great spiritual source. Similar individuals have appeared elsewhere in history, and we shall be meeting some of them in this book, but in my view the Prophet was the last. Years

After my initial brush with Islam I soon started to recognize certain similarities with the other great religions. In particular, they had all apparently been set in motion by single individuals, then, incredibly, had subsequently inspired the voluntary participation, over thousands of years, of millions, billions of people.

So eventually it became clear to me that there are very real forces profoundly affecting the human brain at work within these religious and philosophical movements. Think of Christianity, Judaism, Buddhism, Hinduism, Zoroastrianism, Islam, and so on. Nowhere in the entire "civilized" world is it possible to avoid some kind of contact with one or other of these apparently incomprehensible influences. They emanate from every church, mosque, synagogue, and temple.

Now, recognizing the existence of such forces is one thing, but understanding how and why they operate so effectively is quite another. I pondered over this unfathomable mystery for years, reasoning that the founders of the major religions must have had one fundamental factor in common, which enabled each of them dramatically to affect the lives of whole races of people—but what this was, I had not the least idea.

A BREAKTHROUGH

Then in France, some time in the mid-seventies, a fellow traveler called John Mullins told me about a book he had read recently which had impressed him very much. I asked my wife to send me a copy from England. It was a propitious move. The book in question, *In Search of the Miraculous,* was an account by the Russian writer Pyotr Demianovitch Ouspensky, of his meetings with George Gurdjieff, a Greek-Armenian teacher of "esoteric wisdom" whom he met in Moscow in 1915.

Ostensibly the book is a record of talks given by Gurdjieff to his pupils over a period of about eight years. I had never heard of Gurdjieff or his principal pupil prior to this, but after reading Ouspensky's brilliant piece of reportage from cover to cover, stopping only to eat, drink, and catnap, I can truthfully say that encountering the teachings of this man was one of the most important stages in my entire voyage of discovery. I could write a book on this book, but that would be a digression—and in any case I would simply be diluting what is easily obtainable from any good bookshop. The main thrust of Gurdjieff's teaching, however, I will briefly mention here, because it is relevant to this part of my tale.

Basically he taught that the universe and everything within it is made up of vibrations, resonating, interactive "signals," which permeate through all kinds, aspects, and densities of matter. This almost immediately struck a familiar chord in me, because it reminded me of my earlier impression that all matter is made up of sparkling (vibrating) particles of light. What really made me sit up, however, was Gurdjieff's explanation of how these vibrations move through matter, time, and space.

Gurdjieff said that all processes, all "vibrations," both in the world and in man, are governed by two fundamental laws—laws that were understood in the remotest antiquity.

The first is the law of three, which says that every action, every phenomenon in the universe, is the direct result of the mutual interaction of three forces: active, passive, and neutral. If you ever do something so basic as change a three-pin plug, or use a catalyst in a chemical experiment involving two other compounds, or watch a referee do his job, or examine the structure of an atom, you will recognize immediately the action of these three forces. They are fundamental, everywhere; quite literally, universal. If you have only two forces—active and passive—the result is either deadlock or destruction, but if a third, reconciliatory force is introduced, anything and everything can happen. Gurdjieff said that this concept was the basis of the Holy Trinity of Christian tradition. This in turn implies, of course, that Christianity itself was formulated by people of a scientific turn of mind, people who understood the principle of the three interacting forces, the forces of creation. And, of course, the Trinity, in one way or another, is a fundamental component of virtually every major religion, a fact that suggests that "science" itself—the science of creation—is indeed rooted in the distant past. Scholars may argue that the Trinity was in fact denied by certain monotheistic religions such as Judaism or Islam, which assert emphatically that there is only one God.

But consider this: the most significant act of creation in the whole of Islamic tradition was the revelation to Muhammad of "God's words," which were subsequently compiled as the book known as the Koran. We thus have two participants, Allah and His Prophet. We should note, however, that Muhammad is said to have received his revelations not directly from God Himself, but through an intermediary, the archangel Gabriel—enter the third force. This exact principle is described in the first verse of the first chapter of Genesis: "In the beginning God created the heaven and the earth." You can't get much clearer than that. Creation is the result of three forces. This is the first law.

The second ancient law is the law of octaves. This says that all vibrations moving through matter, and through man, develop—that is, ascend, descend, grow stronger, weaker, and so forth—precisely as a musical octave develops, that is, in proportional steps of seven or eight. Now, this development, apparently, does not proceed uniformly, in a smooth ascension or descension, but erratically, with certain regular "glitches" in the line of motion. Just like a musical octave, in fact.

For the benefit of those unfamiliar with the structure of the octave, or the major musical scale, the notes Do, re, mi, fa, so, la, ti, Do are each separated by a series of intervals or tones, five of which are whole, and two of which are only half-tones, like so:

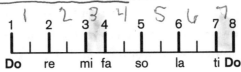

or, as illustrated by the keys of a piano:

The half-tones—between the notes mi–fa and ti–Do—are the glitches just mentioned. Said Gurdjieff, "The seven-tone scale is the formula of a cosmic law which was worked out by ancient schools and applied to music."[1]

Exactly how ancient these schools were is never specified, but Gurdjieff in his own writings suggests that the Pythagoreans, usually credited as being the originators of practical musical theory, had actually rediscovered a long-lost science.

During his talks, Gurdjieff gives Ouspensky many examples in nature demonstrating the action of the forces described by the two fundamental laws. These forces, for example, dictate the structure of white light, the seven colors of the spectrum, and the sevenfold symmetries of the periodic table of chemical elements. He also describes the physical structure of the universe in musical terms, even including in his unique worldview the biological and psychological composition of the human being. All of these phenomena, Gurdjieff said, are essentially musically structured. His evidence, as presented to Ouspensky, was for me extremely compelling, and I instinctively felt that here at last was a real nugget of spiritual knowledge, a genuine, 24-carat kernel of truth. And I was right, as we shall see.

Of course, knowledge is one thing, but understanding how best to use it is quite another. This is what made Gurdjieff unique among all the people whose ideas I had previously encountered, because he didn't simply present original and interesting knowledge, he applied it in an entirely practical and comprehensible way.

I must try to keep things simple at this stage, as my preliminary story is not yet finished. All one needs to know here is that the practical aspect of this knowledge—the core of which is musical theory—is based on a systematic application of these "musical" rules as something like a code of personal conduct. The theory is that, by doing this, by introducing musical rhythms and elements into our lives in an orderly and disciplined way, it is possible for us to evolve, to become more and more conscious (i.e., harmonious) at a much faster rate than is normally envisaged by evolutionists. We can call this the principle of "transcendental evolution," which holds that a harmonious individual is like a fully evolved octave and is capable, through the final "note" Do at the top of the given scale, of striking a single new note, or impulse, into a greater scale above. This greater scale, or dimension, the ancients called heaven. Darwinists take note: what is being implied here is that there are certain limitations to current evolutionary theory, that it is, at best, incomplete.

In between life's periodic distractions, I studied Gurdjieff's ideas for several years, on and off. I read everything by and about him that I could find. He apparently drew his ideas from a number of ancient traditions, as referred to in Ouspensky's book—Egyptian, Christian, Buddhist, Dervish, Hindu, and so on. Of course, these were, broadly speaking, the very same religious movements that had intrigued me for years, which I had surmised were sources of strange metaphysical power that could quite literally move legions of the faithful.

There was an obvious and important link here, but for some time its real significance escaped me.

In the early eighties life slowed down. We had two young children by then and, as there were inevitably a number of conspicuous gaps in my CV, finding a regular and amenable occupation proved difficult. These were days of high unemployment, and the corporate drifter found himself up against stiffer opposition than he had expected. In the end, after several halfhearted attempts to reenter the professional workplace, he decided to go back to school, taking exploratory extramural courses in numerous and often tedious and uninspiring subjects. There was, however, one short course—a module dealing with the biomolecular world, with DNA and the genetic code—that ultimately turned out to be exactly what I had been looking for.

This was in the summer of 1984. I had been reading a great deal at this time, both textbook stuff and books of my own choosing. One week, genetics or astronomy, perhaps, or a droning essay on Karl Popper; another week, John Michell, Colin Wilson, or Idris Shah, or

possibly a couple of chapters of Gurdjieff's monumental epic _Beelzebub's_ *Ref*
Tales to His Grandson. This multilayered tome is over a thousand pages
long and no easy read, with sentences the length of paragraphs, paragraphs pages long, and dozens of obscure new words invented by the
author—for the express purpose, one suspects, of making reading it
an even more difficult task. Gurdjieff advised his followers to read it
three times (presumably in accordance with the law of three). The fact
that I undertook this daunting task may say something about me that I
wouldn't care to hear, but I completed it, nevertheless, in several stages,
over a period of about five years.

I need not comment on the book itself. As with Ouspensky's, I
could write a lengthy treatise on it—or try to—and even then would
possibly succeed in conveying only a small fraction of its intended
meaning. The point I want to make here is that Gurdjieff's ground-
breaking ideas were well to the fore in my mind. In fact, practically all
of the thoughts and ideas I have mentioned so far were jiggling around
in my head, like ephemeral, dancing genes: life . . . death . . . light . . .
timelessness . . . matter . . . vibrations . . . religion . . . force . . . Gurdjieff
. . . music.

It was virtually all there, like the scattered pieces of a jigsaw, but
the overall picture still eluded me.

Now let's return to the genetic code. Probably most of you will at
least have heard of this chemical arrangement, used by the DNA in the
cells of your body to manufacture amino acids, the building blocks of
all organic life.

In order to give myself a kind of visual aid, an image of the code
in action, I had drawn up a diagram incorporating the key numbers
of the biochemical components involved in the process. These were
4, 3, 64, and 22. That is, there are four kinds of chemical bases. It
takes three of them to make what is known as a triplet codon, an
amino acid template, of which there are exactly sixty-four variations. Each of these codons correspond to one or another of twenty-
two more complex components, namely, the twenty amino acids and

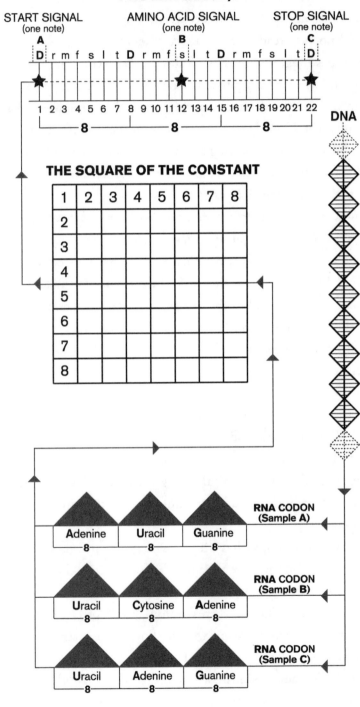

the two coded instructions for starting and stopping the process of synthesis. In my diagram, the number 64, the number of triplet-codon combinations (4 × 4 × 4), was represented by a square grid, eight divisions across and eight down, like a chessboard.

It struck me immediately that there was a curious kind of symmetry involved with these particular number combinations, one that was entirely familiar to me and that I had seen many times before. I realized, in fact, that the whole diagram echoed the format of the famous Chinese work known as the I Ching (Yi King), whose sixty-four basic texts are each identified with a six-line symbol called a hexagram.

The I Ching, the so-called Book of Changes, was one of the more popular works circulating among "New-agers" in the sixties and seventies, and I had browsed through it several times. It was intended for use as an oracle: you pose a question, toss three coins, and note the way they fall. A preponderance of heads gives an unbroken line—"yang," positive; tails a broken line—"yin," negative. There is an older method of consultation using a clutch of forty-nine yarrow stalks, but the principle is the same. Repeat the action six times and you will have called up one of the hexagrams. The accompanying text supplies your answer. Although I was never drawn to actually consult the I Ching, I had always been intrigued by its structure.

As I said earlier, the key numbers of my genetic diagram were 4, 3, 64, and 22.

Let's begin with the number 4, the number of fundamental chemical bases in the genetic code (adenine, thymine, guanine, and cytosine) upon which the whole process of amino-acid synthesis depends. The I Ching, I discovered, embodies exactly the same principle. The sixty-four hexagrams are actually constructed from four, basic, two-line symbols known as the Hsiang. These in turn were derived from the two fundamental lines, one broken and one unbroken, known respectively as yin and yang.

Next, the number 3. The genetic code, as was evident, obeyed the law of three forces, which is why only triplet codons are evident in the process of creation. The three forces are initially represented in the Book of Changes by the two original yin and yang lines, and a third factor called the Great Extreme: that is yin (negative, female), yang (positive, male) and neutral, the third, invisible or "mystical" ingredient, the tao. This greater trinity is fundamental to the whole system, but the number 3 also occurs in a way that corresponds exactly with the genetic code, because each of the hexagrams is described as being composed of two "trigrams," two three-line signals.

We now come to the number 64. As we have noted, the I Ching is composed of sixty-four hexagrams. At first glance it seems as if the genetic code deviates from this pattern, with its sixty-four triplet units. However, it should be noted that the genetic code functions as a dynamic system, and as such should be viewed as an ongoing, evolutionary process, in which every part is connected both with the simpler processes below and also with the more complex components above. Thus we can see that there is, in fact, another side to the codon template, the amino acid itself, which must also, by its very nature, be tripart in structure. So we have one triplet codon and one amino acid—a biochemical "hexagram." Incidentally, triplet-codon templates originate inside the DNA molecule, as copies of segments of its internal structure; this means, of course, that DNA itself is also composed of sixty-four biochemical hexagrams.

By this time, having recognized so many similarities between the I-Ching and the genetic code, I was convinced that I was on to something of profound importance, and my emotional state reflected this: I was highly charged. No way, I thought, could the identical features of these two apparently disparate systems be the product of mere coincidence, for they were not only identical in structure; it seemed that they each had a common purpose, which was to facilitate the process of evolution. Just think about this for a moment: the genetic code is used to create a greater organic structure; the I Ching, the Book of Changes, is supposedly used to create a greater, more enlightened being. The principle is exactly the same.

With a greatly increased respect for it, I returned to the I Ching several times—not to read it or to consult it, but to concentrate on its structure. I felt that its real secrets must lie in the symbolism expressed in its format and that the accompanying texts were simply an embellishment, merely repeating, in longhand, what the hexagrams were already telling me.

Now these hexagrams, as I said earlier, just like the biochemical hexagrams of the genetic code, each consist of two trigrams, two three-line symbols, one above, one below. The trigrams, eight in number, were derived from the four Hsiang, by successively placing over each of them the two original broken and unbroken lines. When these same two lines are placed over the eight trigrams, the result is sixteen figures of four lines. Repeat the process once again and you get thirty-two figures of five lines, and a final similar movement produces the sixty-four hexagrams.

Unlike the four- and five-line figures, the eight trigrams, known as the kwa, are given particular prominence in the system. I mused over these for a long time, juggling with their numbers. Eight threes. Three eights. Twenty-four. I needed twenty-two. Close, but not close enough. Certainly the number 8 was an integral part of the overall symmetry, being the square root of that magical 64; but why did the sum of the trigrams not conform to the twenty-two codon signals of the genetic code? Why twenty-four? Why eight?

It was an exhilarating moment when the light finally dawned and the answer, which came filtering through in the form of the tiniest of thoughts, exploded silently inside my head: "Heptaparaparshinokh."

This peculiar word is one of Gurdjieff's creations, and it is repeated many times in *Beelzebub's Tales*. It means, quite simply, the law of octaves, the law of seven (sometimes expressed as the law of eight),[2] the law by which, he had said, everything proceeds. Everything? Including the genetic code and the I Ching?

So there it was; obviously the symmetry I had been looking at was musically based. It had to be. Here was my chessboard: eight divisions across, like an octave; eight divisions down, another octave; and sixty-four divisions across the grid, an octave squared.

I then remembered what Gurdjieff had written in *Beelzebub's Tales* about the origins of musical theory. He said that the Greeks only redis-covered the science, and that its true origins were far more remote in time. No dates are given, but what he had to say about its originators turned out to be extremely pertinent. Beelzebub informs us that, a very long time ago, there once lived two brothers—princes—in ancient China. These men were direct descendants of a high initiate who sur-vived the cataclysm that destroyed ancient Atlantis, and it was from his teachings, passed down through the generations, that they learned of the law of octaves.

According to most commentators on ancient Chinese history, the creator of the trigrams was a legendary sage called Fu-hsi, thought to have lived in the third millennium BCE. King Wen of the Chou dynasty and his son, Tan, the Duke of Chou, added the texts much later, around 1140 BCE. Princes, kings, dukes . . . it all sounded very familiar.

In the same section of *Beelzebub's Tales* there is a detailed account of how these ancient men of genius verified for themselves the law of octaves (aided by experiments with light, prisms, and other, strange paraphernalia), and how, subsequently, this knowledge became lost. As I recalled how the I Ching was being so casually used simply as a pocket fortune-teller, I could see how true this was. These people never acknowledged the "music" inherent in the system. They "played" it without even knowing.

But now I felt that I had found it again, the secret of life, no less, the music of life, the music in you and in me, in the I Ching and the genetic code—and even, if Gurdjieff's claims hold true, in the cosmos itself.

And the number 22? It fits perfectly, as can be seen from the Pythagorean version of this ancient science. This number was one of the key numbers of their system, principally because of its musical aspect. It represented, in fact, three octaves of vibrations, or notes, three sets of eight—twenty-four components.

If you look at a twenty-two note scale in diagrammatic form, you will see that the first octave is made up of the eight familiar fundamental notes: Do, re, mi, fa, so, la, ti, Do. The eighth note, Do, however, is also simultaneously the first note of the second octave. So the two octaves overlap. Similarly, the eighth note of the second octave—again Do—is also the first note of the third; so these again overlap. In this way we see that the twenty-two divisions actually represent what is, in reality, a manifestation of twenty-four interrelated components—three individual octaves, or 8–8–8:

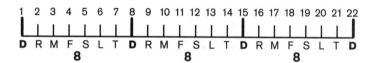

Now see what happens when we apply to the phenomenon described above the first law of nature—the law of three forces—which tells us that each of the individual octaves are themselves tripart in structure, composed inwardly of three octaves each. This produces nine subordinate octaves:

Nine octaves, of course—from the first base note, Do, to the last— contain precisely sixty-four fundamental notes.

INTERIM

The story I have just related is only the beginning of the next, which even now is still unfolding. From that time onward, the summer of 1984, I spent several years exploring the mazelike annals of history. I automatically assumed that, if the Chinese and the Greeks were "tuned in" to this ancient science referred to so frequently by Gurdjieff, then it was probable that so were some of the other traditions and civilizations he mentioned. As it turned out, the evidence was overwhelming. Everywhere I looked I saw musical symbols beaming back at me: every known major religion and esoteric tradition in recorded history had embraced this science as a code of conduct, a harmonious mode of being. Here, in fact, was the missing common factor I had long felt existed, that magical ingredient that had given religious movements the power to affect the minds and hearts of billions in such a profound and extraordinary way. They were all unerringly based on the principle of harmony, a harmony that is echoed in, literally, every single cell of our bodies, in our DNA and in the genetic code. This is, therefore, a natural harmony, one that must naturally appeal to the deepest and innermost instincts of DNA's ultimate creation: *Homo sapiens sapiens*.

Remember, DNA has been successfully using this method of evolution for billions of years. And now look what it is capable of. What began in the primeval soup with a single-celled organism has culminated in the creation of the conscious human being. That's quite a leap, even if it did span four thousand million years.

And then, somewhere along the line—very recently by evolutionary standards—a group of extremely gifted individuals somehow came to realize that the best way forward was to get right back down to basics, to a musical mode of being that was in harmony with the natural evolutionary processes of nature. This, surely, is Science with a capital "s"; it is the science, and in one way or another, it touches all others. And, as we shall see, in terms of the cumulative effect it has had upon the human race, and of the numbers of people who, over several thousand years, have been drawn to study its principles, it genuinely has no peer.

Now, these ancient men of genius, the first practitioners of the noble art of right living, subsequently disseminated their superior knowledge far and wide, across the entire globe, across millennia of time. The results of my researches into this extraordinary cultural phenomenon were finally published in 1994 in my first book, *The Infinite Harmony*.

The book begins with Old Kingdom Egypt, where the symbol of the octave first appeared in the form of their pantheon of eight gods, four male (yang) and four female (yin), said to have materialized on the fabled Island of Flame, the primary source of light. There were, in fact, three coexistent creation myths in Old Kingdom Egypt, and in all of them the octave format is paramount. Furthermore, as with virtually every major religion, Egyptian theology embraced a trinity of three major deities: Osiris, Isis, and Horus—an expression of the law of three, and the triple octave, composed of twenty-two notes. In fact, the mathematical convention *pi (p)*, expressed numerically as 22/7, is first and foremost a symbol of the triple octave, an encoded description of the law of octaves and the law of three forces. Orthodox historians will tell you that this formula wasn't known in Old Kingdom Egypt, but this, as I have shown elsewhere, is an entirely false assumption. Indeed, the *pi* formula was not only known by these people, it was applied in a wholly practical way in respect of their day-to-day activities, and can be identified quite clearly in important administrative documents of the Old Kingdom.[3] This is quite apart from the evidence of the Great Pyramid itself, of course, the structural proportions of which accurately express the mathematical value of "classical" *pi*.

Incidentally, the *pi* relationship has also been discovered in the dimensions and proportions of the so-called Pyramid of the Sun at Teotihuacán in Mexico and, most recently, in the structures of Stonehenge in England and the "step pyramid" of Silbury Hill in southern England—details that suggest that the dissemination of the knowledge of this musical code was a genuine worldwide movement. Later on we can look at these relationships in more detail.

Subsequent sections of *The Infinite Harmony* are devoted to tracing the ongoing development of this musical influence, which flows like a river of pure thought through all the major belief systems in history, including Judaism, Zoroastrianism, Hinduism, Jainism, Buddhism, Confucianism, Christianity, Islam, and the alchemical schools of the Middle Ages. But then, as if this extraordinarily potent stream of ideas flowing unhindered through earlier times were not a remarkable enough phenomenon, later in this book we shall note how the main tenets of this ancient teaching are now finding conceptual niches within the framework of the most advanced, systematic disciplines of our own age. We see this not only in the science of genetics, as just discussed, but also, as will become apparent, in particle physics and even astrophysics and cosmology.

The final chapter of my first book takes a detailed look at the musical structures of the biomolecular world, the realm of DNA and the genetic code, of amino acids and protein macromolecules.

Now this musical symmetry is real, it exists for all to see, and I don't believe I have contrived it in any way. I have merely looked at the facts as presented to me by experts in their respective fields, and then strung them together into what I see as a clearly recognizable musical pattern. Music is life; life is music: this is what I have learned from the facts. But of course the most important, and perhaps controversial, fact of all those arising from my research is that the musical symmetry dictating the evolution of the human gene pool was not only identified by ancient peoples, it was actively employed in their daily lives as a complete way of being, a "religion."

As the first recorded version of this archaic science first appeared in the Nile delta about five thousand years ago, I have called this musical symmetry the Hermetic Code, after Hermes Trismegistus, the Greek name for Thoth, the ancient Egyptian god of wisdom. This musical symmetry, as I have said, is precisely what the formula *pi* was designed to express; that is, the law of seven and the law of three—the triple octave, composed of twenty-two notes. And remember, the law of three tells us that each of the individual octaves so expressed is also tripart in structure: it is composed of three "inner" octaves, making nine octaves in total—exactly sixty-four fundamental notes, the square of the constant number, 8. This is the Hermetic Code, a universal formula that, as we shall find out, encompasses within it practically everything.

Having thus outlined my preliminary ideas, it is now time for us to embark on a greater journey of discovery—to new territories that I myself have yet to explore fully. Therefore much of the discussion that follows is to a large extent speculative in nature. We shall be dealing with "facts," of course, many of them astonishing—unbelievable even—but empirical data, however pertinent, can take us only so far. If we wish to go to realms existing beyond the bounds of "logical" thought—which is where this narrative is intended to take the reader— then we may have to temper our "knowledgeable" worldview with liberal measures of two of the most elusive of human faculties: instinct and intuition.

Instinct, as most people understand it, is a gut feeling, something in one's bones. I am sure that we have all experienced this impulse in one way or another. Instinct may well be the primary cause of the emergence, over the past five millennia, of the inordinately powerful religious movements referred to earlier. These are long-held traditions, essentially rhythmic in form and method, and they are steadfastly adhered to by billions of ordinary people even now. These religions were founded, as I have explained, on the principles of musical symmetry, a symmetry that is so clearly evident in our DNA—literally in our bones. This could mean that the inclination to pray, for example, on every seventh day may, to a large extent, be the product of instinct. The "holy" Sabbath is the seventh day/note in an octave of time; the eighth "note"—Sunday in Christianity—is also the first day/note in the next "octave," the next week. Clearly this essentially "religious" division of time is no arbitrary invention. It is musically based. It has "rhythm," it has "soul"; it is a perfect example of real "live music," of the Hermetic Code in action.

Now intuition, on the other hand, is generally believed to operate from what we call our subconscious minds—a seemingly inexplicable, faster-than-light process of instantaneous recognition, flashing on and off in sparse, random bursts, conveniently providing us with sensible answers to "impossible" questions—and very often when we least expect them. If this has never happened to you, the idea might seem too vague or fanciful to be taken seriously. To dispel any doubt, consult the recognized experts, by which I mean trained and disciplined thinkers in virtually all fields of scientific investigation, past and present. I think you would be hard-pressed to find more than a handful of these individuals working today who would deny the existence of the intuitive process in the functioning of the questing human mind. Indeed, as we shall see later, modern scientific enquiry thrives on it.

To put all this simply, I am asking readers to open their minds and try to put aside all preconceptions about, well, almost everything. I realize, of course, that this is a tall order in these troubled and confused times, but the expansive, sometimes dizzying journey we are about to make requires that we be, as it were, "fleet of foot," relatively free from dogma and conceptually ready for anything. Having said all that, there is one crucial fact that we need bear in mind: that the Hermetic Code, the blueprint of all creation, has been known and understood as such since the dawn of our civilization.

As will become clear, the originators of this code discovered a spectacular sphere of existence, its expanse far beyond the confines of today's imagination. It stretches backward and forward in time, to infinity: through historical time (as far as it goes), through geological time, to the first stirrings of life on earth—and even further still, through cosmological time, to the origin of the universe itself, the big bang. This remarkable worldview also encompasses all kinds of space: the inner space of the molecule, the atom and the subatomic particle, the space you and I perceive as "real" space, and the greater scales of space above us—the space of the planet, the solar system, the galaxy, the galactic supercluster. In truth, there is not a single phenomenon that is excluded from this all-encompassing cosmic plan, not even life, death, and the ultimate fate of our expanding universe.

Obviously, therefore, at certain stages in this investigation we shall have to look at some of the weird and wonderful notions of modern science—in particular, some of the key advances made in particle physics, astrophysics, genetics, and evolutionary theory. As a nonscientist,

I trust that my interpretations of these complex, sometimes perplexing, ideas, will be clear enough for most readers to follow. In any event, they are necessary excursions, as a basic understanding of recent scientific thought will enable us to compare it with some of the theories of ancient times. There are a number of surprises in store, for today's science appears in many ways to be simply reiterating what has gone before. This is not, of course, what scientists themselves want to hear, and, as a blatant trespasser, I would not expect them to give these ideas a warm reception.

So, where do we begin such an apparently impossible journey across an infinity of time frames and even dimensions? It is probably best to start with our feet firmly on the ground—on bedrock—in the land and time of Old Kingdom Egypt. The reason is clear: because the *pi* convention first emerged here, both in the architecture and in written form.

1
The Sacred Constant
The "Jewel in the Crown"

Although all ancient civilizations were special in their own way, Egypt was in a class entirely of its own. To begin with, it was the first truly unified nation in history. It was also the longest-lived, spanning three thousand years from unification to final dissolution. Its architecture is truly exceptional: in magnitude, sophistication and precision, nowhere since has it been surpassed or even equaled. Most importantly, it was in Egypt, at the very beginning of the era of the historical pharaohs, that the Hermetic Code first came to light. This is demonstrated not only through the *pi* relationship, which appears to have been incorporated both in the architecture and in the administrative procedures of this culture, but also in its protracted and detailed mythology. The octave format inherent in the natural processes of creation was first expressed in stories surrounding the miraculous appearance of the eight principal gods of the Egyptian pantheon.

It was once pointed out to me that one of the earliest and most influential of the Egyptian creation myths portrayed a pantheon not of eight but of nine gods, known in the Old Kingdom as the Great Ennead. Clearly this rather awkward detail was disconcertingly inconsistent with "my" musical theory. The obvious anomaly puzzled me at first, but on closer examination of the myth in question, I came to realize that there was, in fact, no inconsistency whatever. More than that, I found that this imagery of a mythical group of nine not only embodied within it the symbol of the octave, the primary ingredient of the Hermetic Code, but also an extremely subtle connection with another key component of this universal formula.

The principal god of the Great Ennead, Atum, was said to have fertilized himself to produce eight offspring. So we are already back on safe ground. Eight is good. So is seven, of course, as expressed in the formula *pi*, the symbol of the triple octave, the "trinity." In general, the numbers 7 and 8, as they appear in myth and religious tradition, each refer to the same concept, namely, the octave and the musical symmetry inherent in all processes in nature. And the number 9? This also has definite musical connotations. If we substitute the "gods" of the Great Ennead with the word *octaves,* the result is a musical composition of nine octaves, comprising sixty-four fundamental notes. The Great Ennead, therefore, is simply a mythical expression of the formula *pi,* the triple octave, which is quite naturally subdivisible into nine inner octaves, sixty-four notes.

Sixty-four is, of course, the square of eight. And the number 8, as I have said, is a constant number, one that consistently recurs in nature, in the genetic code, in the spectrum composing white light, and in the natural harmonies of the major musical scale. The Egyptians evidently recognized this musical symmetry, which is why they held the number 8 to be sacred—hence its association with the gods. Eight was thus regarded as a "sacred constant," a yardstick by which everything could be measured or compared. Sixty-four, being the square of this number, was therefore of supreme significance to the guardians of the Egyptian mysteries, representing the ultimate goal of the individual—the squaring of one's possibilities, the acquisition of godlike attributes. In "bioharmonic" terms we might call this the attainment within oneself of an optimum degree of physical and psychological "resonance," an individual condition of absolute metaphysical harmony. This is simply a higher state of consciousness, a level of perception that empowers the individual to strike metaphysical or conceptual "notes" up in a greater "scale" above, which we call, for want of a better word, heaven.

ORIGINS

I have often speculated on the true age of this ancient wisdom. As far as my own research has revealed, it seems initially to have appeared in its full-blown form in Egypt in the first half of the third millennium BCE. This by no means proves, however, that the concept of the sacred constant actually originated in Old Kingdom Egypt.

Even when starting to write *The Infinite Harmony,* I suspected that the canon of wisdom to which I had tuned in could be much older than the existing records show. Orthodox historians are reluctant to push back the beginning of Egyptian civilization much further than the establishment of what is known as the Archaic Period, which began around 3100 BCE, when Upper and Lower Egypt were first unified under the rule of King Narmer and his successor, Hor-aha, or Menes. However, the Egyptian chronicler Manetho, a priest of the city of Heliopolis from the third century BCE, recorded a long, continuous succession of divine and semidivine rulers of prehistoric Egypt stretching back 24,925 years beyond the beginning of the Archaic Period.[1] An earlier, fragmented document known as the Royal Canon of Turin and dated to around 1300 BCE contains a king list that begins with an unspecified period when Egypt was ruled by a succession of ten gods, the Netjeru, followed by a second period of 13,420 years of divine rulers known as the "Followers of Horus."[2] Obviously these reigns are given no credence by academics: the timescales involved are simply too great to fit into any accepted historical format. But in my view these records, though they might have become distorted with the passage of time, are highly significant, primarily because they reflect the views and traditions of the Egyptians themselves. Clearly these people firmly believed that their culture had its origins in a past reaching back many thousands of years before the Archaic Period.

So which view is correct? Is it that of today's archaeologists and historians, who rely principally on the evidence of datable artifacts to ascertain the age of a culture? Or is it the account that has come down to us from the horse's mouth, so to speak, from the Egyptian priesthood of Manetho's day and from the compilers of the Royal Canon of Turin?

It seems to me very unlikely that a culture as advanced and sophisticated as Egypt's should have mushroomed "overnight" from a primitive intellectual environment. As we have noted, the Old Kingdom priests and astronomers were already in possession of a fully developed, extraordinarily imaginative belief system—a true science, no less—the main tenets of which they described symbolically and with superlative clarity in the form of the Hermetic Code. What is, perhaps, even more incredible is the obvious intent behind this essentially harmonious mode of spiritual evolution, which was to forge for mankind a direct means of access to the stars, to the mythical realm of the gods that came to be known as heaven. The important point to note here is that heaven—that place I vainly looked for as a boy—is, in fact, a scientific concept employed by ancient metaphysicians to describe a higher dimension of existence. We can define this netherworld in musical terms as a greater "scale" of being. This unique mode of evolution is based, as I have said, on systematic scientific principles—those of harmonics—but later we shall see that this theory of "transcendental evolution" is also scientific in other significant ways—specifically in relation to some of the benchmark discoveries of modern physics and genetics. But even if we presently consider only the musical aspect of this theory, this is a surprisingly sophisticated concept to have originated with the relatively close descendants of simple hunter-gatherers.

In addition to this highly evolved "religion" of the Fourth Dynasty Egyptians, the stonemasons and architects of that unique era—particularly the builders of the Pyramids of Giza and Dashur—displayed incomparable expertise in precision and enormity of scale.

So history here belies logic: the finest, the most sophisticated, appeared very early on, and the standard of pyramid building degenerated as time went by. This is not, of course, what we observe in the ongoing development of our own culture, particularly in technology and the sciences, which are generally considered as having progressively evolved—primarily out of the supposedly superstitious ideas of medieval alchemy.

Of course, if the Egyptians themselves are to be believed, quite the opposite appears to have happened. According to them, their culture was not the product of a gradual development, but began in full bloom, with a first, golden era, a distant age when the "gods" suddenly appeared (presumably from somewhere), bringing with them all the trappings of civilized existence. This period in their history the Egyptians called *Zep Tepi,* the "First Time."

From this golden age onward, there apparently proceeded a gradual process of involution, through a long period of high civilization under the rule of lesser demigods—the "Followers of Horus" earlier mentioned—ultimately ending with the "normal" era of the historical pharaohs. I use the term *normal* guardedly here, because the Great Pyramid was constructed at this time, and it is difficult to imagine that those responsible for the creation of this amazing structure were just ordinary souls.

On the face of things, it seems as if this ancient Egyptian scenario of a distant, perfect beginning is, by conventional standards, fantastic, totally at odds with the established academic view of events. Egyptologists tell us that the datable, factual evidence, painstakingly collated by scholars over the last couple of hundred years, proves conclusively that a mere one thousand years before Egypt's Dynastic Period, the tribes in the Nile region were living in simple, mud-brick dwellings and thought about little other than survival.

And then, one thousand years later—around 2550 BCE—historians present us with the fully developed civilization we know as Old Kingdom Egypt, the most advanced of the ancient world, whose architects built like giants, and whose influence upon the human race is felt even to this day. This crucial development, from the hunter-gatherer tending a few goats in the middle of the fourth millennium BCE, to the Egyptian priest-astronomer of the Old Kingdom, with their gigantic pyramid markers and lofty thoughts of the firmament, is perhaps the most dramatic leap forward ever encountered in human history.

This sudden appearance of the high civilization exemplified by the architecture and the extant texts of Old Kingdom Egypt has prompted many alternative investigators to challenge seriously the orthodox view of the sources of Egyptian culture, positing a direct legacy from an earlier, greatly more advanced race of people.

The evidence in support of this revolutionary proposition has in recent years come to us from all quarters, and doubtless most readers hooked on the mysteries will already be familiar with most of it. Modern writers, such as Graham Hancock and Robert Bauval, Colin Wilson, Andrew Collins, and many more have done much to increase public interest in "alternative" explanations of the origin of civilization. Many of the ideas discussed by these authors are of necessity highly contentious and so are continually under attack from academics. However, even if we accept that some of these theories are, to coin an apt phrase, not "hermetically sealed"—not entirely watertight—there is now a whole swath of significant new data that simply does not fit in with the orthodox picture of events. In short, it is now time for us to rewrite the standard, but apparently garbled, story of the development of civilization, for the period that historians have consistently referred to as "prehistory," circa 10,500–4000 BCE, is that no more.

Most of the new evidence to support this view of a highly advanced, proto-Egyptian culture has been variously dealt with by dozens of alternative theorists, so we need not dwell too long on the details. A quick review of the main features will suffice to impress upon the reader how remarkably advanced these denizens of "prehistory" really were. No doubt academic arguments over the true age of civilization will rage for years to come, but this is not my primary concern. The intention of this book is to proffer an alternative perspective that is not so much focused on the antiquity of our culture, but more on the nature and future ramifications of the unique belief system bequeathed to us by our most distant ancestors. But first let us examine certain key, well-established facts and see for ourselves exactly what these mysterious people were capable of.

"IMPOSSIBLE" MAPS

In 1966, the late Charles H. Hapgood, Professor of the History of Science at Keene College, New Hampshire, published a seminal work called *Maps of the Ancient Sea Kings*. The title refers to certain medieval maritime maps currently filed away in the American Library of Congress, which depict the exact contours of the land mass buried deep beneath the mile-thick ice packs of Antarctica. Possibly the best known of these is the now famous Piri Re'is map of 1513, owned by a Turkish pirate executed in 1554. This shows the southern Atlantic Ocean, a part of the coast of Africa to the right, the coast of South America to the left, and, farther to the south, Antarctica, with the bays of Queen Maud Land shown in astonishingly clear relief. Note that these bays were not officially "discovered" until surveys using sonar soundings were conducted by a joint team of British, Swedish, and Norwegian scientists in 1949. There are, in fact, many more of these maps—called portolans—some of which show that Antarctica actually consists of two separate land masses, a fact not known to modern geographers until further surveys were conducted in 1958.

Hapgood's explanation for the existence of these portolans remains controversial even today. He surmised that they were probably copies of earlier maps, which in turn might have been copies of even older ones— and so on, effectively reaching back to a time when Antarctica was relatively free of ice. Controversial this proposition may be, but Hapgood's observations were based on solid factual evidence that has never been refuted or otherwise explained by the academic establishment.

The obvious implication of all this is that there may have been highly proficient mariners alive then—when Antarctica was a more temperate land—with the wherewithal to chart it. But of course mapmaking of the accuracy found in many portolans is a highly exacting science and implies a detailed knowledge of trigonometry and geometry—hardly the skills we would normally ascribe to "primitive" peoples.

Geologists and climatologists tell us that the latest possible date that Antarctica could have been a temperate region—and therefore topographically surveyed by ancient seafarers—was around 4000 BCE. If we assume, then, that this prehistoric civilization of accomplished mapmakers and mathematicians progressively evolved, as we have, through long periods of trial and error and spasmodic bouts of inspirational genius, then the conservative date of 4000 BCE would merely mark a turning point in the development of a culture that had already reached a marked stage of maturity. How long it took to attain such a level of sophistication is anybody's guess, but if historians are ever to make sense of the evidence provided by portolans, this is certainly a question that needs to be addressed. Where these ancient seafarers went after Antarctica finally became covered with ice is also crying out for an answer. After all, they must have gone somewhere, because they had boats and were obviously inclined to use them. It is perhaps significant that the time frame under discussion comfortably encompasses the beginning of the Archaic Period of ancient Egypt. And the early Egyptians, in fact, also had boats. Khufu's father, Snefru, had a fleet of them, and they were identical in design to reed boats still being built today by the South American Indians living on the shores of Lake Titicaca in Bolivia. Local Indians informed Graham Hancock that the design had been given to them by the legendary Viracocha, the white god from the sea who, like Osiris in Egypt, brought with him civilization and a new way of being.

In 1954 a rectangular pit was discovered on the south side of the Great Pyramid that contained a dismantled boat made of cedar. It took fourteen years to reassemble this craft, which measures 43 meters from prow to stern.

According to the sailor and explorer Thor Heyerdahl, the streamlined hull of this boat could never have withstood the conditions of the high seas. It was built, in his opinion, as a symbolic craft for the use of the pharaoh in his afterlife. This acknowledged expert in the field of ancient shipbuilding believes that the boat was essentially a riverboat. Curiously, however, he also asserts that its high-prowed design was a highly sophisticated and technically accurate model, not of a riverboat, but of an ocean-going vessel. Heyerdahl is suggesting that Khufu's ceremonial boat could have been derived from the plans of very experienced shipbuilders, people with a long tradition of sailing on the open seas.[3]

It is tempting to think that this is the answer, that the Egyptian and, perhaps, the early American civilizations, whose descendants are still making boats of an identical design today, might have been spawned by descendants of this long-forgotten race of mariners of ancient Antarctica—say, around the middle of the fourth millennium BCE.

The emerging historical picture, however, is not quite that simple.

THE GEOLOGICAL EVIDENCE

Let us now look briefly at the latest controversy surrounding the dating of the Great Sphinx and its two neighboring constructions on the Giza plateau, the extraordinary buildings known today as the Sphinx Temple and the Valley Temple.

In 1979, the American writer John Anthony West published a book, *Serpent in the Sky*, in which he discusses the ideas of the maverick Alsatian Egyptologist, René Schwaller de Lubicz. Schwaller, who spent almost fifteen years between the wars investigating numerous archaeological sites in Egypt, noted marked differences between the erosion of the Sphinx enclosure and that of the nearby tombs and pyramids on the Giza plateau. The weathering pattern on the exterior surfaces of the tombs and pyramids is angular and irregular, with rock layers higher up in the masonry showing less weathering than the layers below. This angular erosion is attributed to the effects of wind-blown sands blasting away the softer layers of limestone and leaving exposed the harder levels. The weathering of the Sphinx and on the outer walls of the two neighboring temples, however, is rounded, undulating, sloping slightly outward toward the ground, with deep, intermittent, vertical fissures or gullies. Schwaller's explanation for this is that the erosion of the Sphinx and its surrounding structures has been caused not by sandblasts, but by water. This clearly suggests that they must have originated in a different and quite distinct era—when water other than that of the River Nile abounded.

West's book subsequently achieved only a moderate success. One can perhaps understand why: it certainly had no chance of being officially endorsed, and the present "Egyptian renaissance" had not yet begun to flower. Undeterred, West, convinced that the weathering pattern of the Sphinx provided a vital clue to our understanding of "prehistory," has spent the last twenty years steadily chipping away at another crumbling edifice—the orthodox opposition to Schwaller's theory. The turning point in this modern David and Goliath story came when West invited an expert to examine the Sphinx erosion and try to ascertain what had caused it. Significantly, this expert was not an Egyptologist, but a scientist, a leading authority on the structure and nature of rocks.

This was the eminent Boston geologist Dr. Robert Schoch, whose studied opinion, based on exhaustive investigations at the site, is that the erosion is indeed the result of water—rainwater, to be precise. Schoch presented his findings at the annual convention of the Geological Society of America in 1992, and his evidence and conclusions were received with marked approval. The assumption is, therefore, that the weathering of the Sphinx enclosure has most likely been produced by precipitation. As with the case of the portolans mentioned previously, orthodox historians may have another serious question to address here, because it has now been fairly well established that Egypt ceased to be a temperate zone nine thousand years ago—around 7000 BCE—over three and a half thousand years before the beginning of the Dynastic Period. Significantly, this ties in rather well with Schoch's estimate for the age of the Sphinx, which he puts conservatively at nine thousand years. If he is off by only a thousand years or so, and the Sphinx was indeed drenched by rain in a temperate climate for a

thousand years or more, this could arguably account for the erosion observed by geologists.

Historians and archaeologists of the old school, however, have reacted negatively to this radical view, insisting that the Sphinx is contemporary with the Dynastic Period. They claim that it was commissioned by, and is an image of, the pharaoh Khafre (Greek, Chephren), builder of the second of the three Pyramids at Giza. An undamaged statue of Khafre was found in the Valley Temple, and the American archaeologist Mark Lehner, leading the academic opposition to West's and Schoch's theories, declared that the similarities between it and the face of the Sphinx proved conclusively that the Sphinx was an image of Khafre. West totally disagreed; in fact, he could see no similarity whatsoever. And neither could leading New York forensic artist Detective Frank Domingo when, at West's behest, he went to Egypt to check out the validity of Lehner's claims. After a close, professional examination of all the available evidence, Domingo's verdict was emphatic: the Great Sphinx is definitely not an image of the Fourth Dynasty pharaoh Khafre. Indeed, if the geologists have it right, how could it possibly be?

If the orthodox view of an evolving culture is correct, we might reasonably assume that the people responsible for such marvelous feats of construction would have had a good deal of practice before they mastered their formidable skills. Therefore if we are considering here a civilization dating back to a time when the Sahara was green, there may be a great deal more evidence of it yet to uncover—a hidden empire under the sand.

MEGALITHIC LEGO

There is also another anomaly intrinsic to the Sphinx enclosure, which again throws the orthodox view into question: the incredible size of the stone blocks used in its construction. Weighing at least 200 tons apiece, with some of them tipping the scale at a staggering 450 tons, these enormous blocks—hundreds of them—have been superimposed with a degree of precision that makes the mind boggle. The point is that these massive, austere, and predominantly rectangular constructions—the so-called Valley Temple of Khafre and the Sphinx Temple—are entirely uncharacteristic of the architecture of the Old Kingdom Pyramids and tombs—the latter incorporating elaborate carvings, inscriptions, cylindrical fluted columns, and numerous other architectural features typical of that era. One possible reason for this anomaly, as a number of investigators have of course already suggested, is that the Sphinx enclosure and the nearby pyramids and tombs of Giza originated in two quite distinct and widely separated time frames.

Take the Great Pyramid, for example, possibly the finest and most complete example of stonemasonry in existence, constructed mainly out of standard blocks of limestone averaging around two to three tons in weight. Even the most massive of its granite blocks, incorporated in what is now commonly known as the King's Chamber, would probably weigh in at "only" around seventy tons.

Now consider this: the average weight of the carved stones used in the building of the Sphinx and Valley Temples at Giza is two hundred tons. Of course, these blocks have been carved from much softer limestone bedrock but, by Old Kingdom standards, they are truly cyclopean. What is more, their joints are so fine that it is impossible to slide a razor-thin blade between them. Even by modern standards, they are nigh on perfect.

In Abydos, in Upper, or Southern, Egypt, there is another ancient structure, a temple known as the Osireion, which has the same stark, megalithic form of architecture as the buildings of the Sphinx enclosure. Professor Edouard Naville, sponsored by the Egypt Exploration Fund, excavated much of the site between 1912 and 1914, and when he observed the unique style of architecture of the structure, he straightaway compared it with the Valley Temple at Giza. Both were made with gigantic blocks without any ornament but, as Naville noted, the Osireion, though of a similar style, was made with even larger blocks, a fact that, rather curiously, prompted him to suggest that it was "of a

still more archaic character."[4] Is it not strange that he should have associated antiquity with size in this way? He was saying, in effect, that the larger the blocks—and, consequently, the more advanced the engineering techniques required to carve, transport, and position them—the greater the antiquity of the building. Naville suggested, in fact, that the Osireion might be the most ancient building in Egypt. This is, of course, a decidedly unorthodox view, but one that accords with that of the Egyptians themselves, with their belief that their civilization began with a godlike race of beings who possessed supernatural powers, in other words extraordinary skills.

Later excavations by Henry Frankfurt at the site in Abydos, between 1925 and 1930, unearthed a cartouche of the Nineteenth Dynasty pharaoh Seti I carved in granite above the entrance into the main hall of the temple. Not surprisingly, this and other minor finds linking Seti with the site prompted the archaeological community to disregard Naville's earlier conclusions, opting for the much more palatable idea that the building dated back to an established period in known history. There is now, however, a growing number of investigators in the field who are inclining more and more toward Naville's view. The writer Andrew Collins, in his book *Gods of Eden*, has suggested that Seti, having recognized the unquestionable sanctity of this important edifice, may have constructed his own temple at Abydos "to comply with the existing orientation and ground-plan of the Osireion, which was already of immense antiquity even in his own age."[5] This makes perfect sense. And, after all, why should Seti, of all of the pharaohs to have ruled in any dynasty since the Fourth (the supposed era of construction of the Valley Temple), be the only one to have built with blocks of such size?

In respect of megalithic architecture, it is worth noting here that the Valley Temple and the Osireion, while unique in Egypt, do, in fact, have counterparts elsewhere in important archaeological sites worldwide, notably in the Lebanon, Bolivia, Peru, and Mexico. In his book *Fingerprints of the Gods* (1995), Graham Hancock describes the awe-inspiring remains of buildings all over the ancient world that have been constructed from massive stone blocks weighing several hundred tons apiece—again, as with the blocks incorporated in the Sphinx Temple and the Osireion, featuring joints of near-perfect precision.

Furthermore, in Mexico, as in Egypt, the pyramid structure is a central feature. Most significantly, in the case of the Pyramid of the Sun in Teotihuacán in Mexico, we even find the crucial *pi* relationship incorporated within its dimensions. This and other "hermetic" buildings will be revisited in due course.

THE ASTRO-ARCHAEOLOGICAL PERSPECTIVE

The fact that the Sphinx has the body of a lion has prompted certain investigative authors to suggest that it was very likely carved in the Age of Leo, some time between 11,380 and 9220 BCE.

To understand the basis of this new argument, we first have to consider the "phenomenon of precession," which is the apparent backward motion of the twelve constellations of the zodiac in relation to the horizon. Precession occurs because the earth is spinning like a giant top that is losing momentum and has begun to turn and wobble very slowly—almost imperceptibly so—in the direction opposite to its spin. This alternative motion gives observers on earth the illusory impression that the "fixed" stars of the firmament are revolving slowly around us like a giant stellar wheel. The rate of precession is measured by observing the gradually changing zodiacal backdrop against which the sun rises on the "spring equinox." Proceeding at the rate of one degree every seventy-two years, this means that each of the twelve ages of the zodiac occupies 30 degrees of this great astronomical circle, taking 2,160 years to complete. A whole precessional cycle therefore lasts 25,920 years.

In our present astrological age, the equinoctial rising of the sun still occurs in Pisces, the symbol of Christianity. As Andrew Collins points out in *Gods of Eden,* just before this age the equinoctial sunrise occurred against the backdrop of Aries, the sign of the ram, which is historically associated with the reign of the Hebrew Scriptures patriarch, Abraham, and also with the ram cult of Amun in ancient Egypt, both of which appeared shortly after this age began, around 2200 BCE. The latter stages of the age before this, the Age of Taurus, the time of the bull cult of the Mediterranean and the Apis cult of Egypt, is the epoch in which Egyptologists say the leonine Sphinx was carved. To Collins, and indeed to many other current investigators, placing the lion squarely in the bull's domain makes no sense at all. The "alternative" view, of course, offers a much more logical explanation: the Sphinx has the body of a lion because it was carved and orientated in the Age of Leo.[1]

The new evidence in favor of this hypothesis is concerned primarily with certain stellar alignments that appear to have been intentionally built into the Giza necropolis. As we might expect, the authors engaged in this new field of enquiry are consistently under attack from traditional Egyptologists, recognized experts in their own field, of course, but scholars who generally know little or nothing about astronomy. Consequently the Leo hypothesis is seen as simply another crank theory being purveyed by ill-informed amateurs.

Possibly one of the best-known names in this new area of "astro-archaeology" is the Belgian construction engineer Robert Bauval. His first book, *The Orion Mystery*, co-authored with the writer and publisher Adrian Gilbert, is based around his discovery of an evident alignment of the Giza necropolis with certain key stars in the Egyptian sky, stars that, significantly, were of very special interest to the priest-astronomers of the Old Kingdom.

Bauval noted the curious misalignment and marked difference in size of the much smaller third Pyramid of Menkaura (Mycerinus), and wondered why this should be. He subsequently realized that this peculiar anomaly corresponded accurately with a similar "misalignment" evident in the positions of the three stars of Orion's Belt, called by astronomers Zeta Orionis, Epsilon, and Delta. The smallest of this triad—Delta—is slightly offset from the line described by the first two. Significantly, it is also noticeably dimmer, apparently smaller.

As Bauval already knew from his exhaustive examinations of many of the so-called Pyramid Texts, the constellation of Orion was extremely important in the mythology and religion of the ancient Egyptians. It was associated with their principal god, the great civilizer, Osiris, Lord of Zep Tepi, the "First Time," the golden epoch when Thoth/Hermes is said to have imparted to mankind the fruits of his infinite wisdom. Orion, therefore, is clearly a key to the enigma of the Pyramids.

Bauval realized further that the Great Pyramid itself has certain internal features that link it directly not only with Orion, the star of Osiris, but also to another equally important star—Sirius—that associated with Osiris's consort, the goddess Isis. Sirius's heliacal rising (i.e., the same time as the sun) was in fact the basis of the Egyptian "sothic" calendar. The link between Orion, Sirius, and the Great Pyramid is to be found in two of the four mysterious shafts projecting from the so-called King's and Queen's Chambers. Using computer-simulated star charts to reproduce the position of the stars above the Nile Delta around 2500 BCE, the time when the Pyramids were built, Bauval established that each of the shafts would then have targeted particular stars as they culminated at the meridian, that is, as they reached their highest point above the horizon. The crucial ones turned out to be the southern shafts. That of the King's Chamber, angled at 45 degrees 14 minutes, would have targeted the star known to the Egyptians as Al Nitak, Zeta Orionis, the lowest of the three stars of Orion's Belt. Similarly the southern shaft of the Queen's Chamber, angled at 39 degrees 30 minutes, would have aligned with the high point of Sirius, or Alpha Canis Major, in the constellation of the Great Dog.

During the course of the precession of the equinoxes, Orion's Belt moves through the heavens in a specific and unchanging way. It rises upward for almost thirteen thousand years, tilting slightly in a clockwise motion, and then back again, drifting slowly down and turning anticlockwise as it returns to its starting point. To Bauval, this starting point was highly significant, for his computer star charts told him that the last time Orion was at its lowest point in the sky was circa 10,450 BCE, in the Age of Leo.

Bauval further noted that at this point in the precessional cycle the three stars of Orion's Belt would not have been tilted sideways, and so would then have perfectly reflected the position and orientation of the three Pyramids of Giza. He surmised that the Giza site in fact acted like a giant star-clock marking the epoch of the First Time, the golden age of Osiris/Orion.

In a later book, *Keeper of Genesis* (1997), jointly written by Bauval and Graham Hancock, the astro-archaeological theory is explored further. They suggest that the designers of the Giza complex saw the River Nile itself as a reflection of the diffused band of light of the Milky Way—our own galaxy. The leonine Sphinx faces due east. Had it been there, as Bauval and Hancock imply, at dawn on the all-important spring equinox in the year 10,450 BCE—the beginning of the present precessional cycle—the constellation of Leo would have appeared above the horizon directly in front of it. Understandably, they see this as compelling evidence that the Sphinx enclosure—which the geologist Robert Schoch believes is much older than the nearby tombs—was built to align with the constellation of the precessional age in which it was carved and constructed. At that time, Orion's Belt itself would have been positioned in the southern sky, at right angles to Leo, and at its lowest point of declination. What is intriguing is the idea that the builders of Old Kingdom Egypt, working in the later Age of Taurus, may have built and aligned their three mighty pyramids to reflect perfectly the position of the three stars of Orion's Belt as they would have appeared in the skies in the Age of Leo. It's as if the designers may have been focusing back on this age for a specific purpose, possibly to mark it as an important era in their history, when Orion was closest to home and when Leo appeared in the sky exactly in line with the Sphinx's present gaze. Orion has been rising in the sky ever since, and will continue to do so until around the year 2550 CE, thus marking the first half of the full precessional cycle, thirteen thousand years after the First Time.

Clearly the earth–stellar configuration being described here, although not yet accepted by academics, has much merit. It is, after all, based on verifiable facts, data that Egyptologists, if they are sincere, simply cannot afford to ignore. The point is that these ancient, highly accomplished construction engineers of Old Kingdom Egypt were also experienced astronomers, people deeply concerned with the movement of the heavens, and they apparently knew about precession. This obviously begs the question, How did they obtain this knowledge? If we assume through a vigilant observation of the heavens, then it must also be accepted that these observations must have continued uninterrupted for a considerable period of time. It takes seventy-two years—a good lifetime—for the zodiacal wheel to move just one degree of arc. So let's say that, somewhere and at some time in the remote past, one rather shrewd individual happened to notice a very slight change in the position of a particular favored star. Fortuitously, he or she then passes this information on to a son or a daughter, or to a group of followers, who subsequently continue to observe the same star. For how long would this observation have to continue before it was realized by someone in the chain that their favorite star had a high point and a low point in its movement through the heavens? More importantly, how long would it be before someone could work out the last time the given star was at its lowest point of declination in a great astronomical cycle spanning almost twenty-six thousand years? (25,920)

So, how old is "civilization"?

In fact, evidence of knowledge of precession in ancient times comes from many quarters, and is not exclusive to the Egyptians. It was mysteriously encoded in the different mythologies of peoples from all over the world. This was first noted by the scientific historian Giorgio de Santillana and the anthropologist Hertha von Dechend, in their complex and challenging book *Hamlet's Mill* (1960).

Ref

Put simply, what de Santillana and von Dechend discovered was that certain numbers—derivatives of the great precessional cycle of 25,920 years—cropped up again and again in myths and legends from cultures all around the world. As with the numbers associated with the laws of harmony and the structure of the octave, they appear to be deeply rooted

see p.39 $25,920 \div 12 = 2160$

in mankind's race-memory. The numbers are all based on the 360-degree precessional wheel ("Hamlet's Mill") and the numbers of years occupied by the twelve zodiacal ages. As we noted, one degree occupies seventy-two years, which is one of the key numbers in the series. Further, one "age," one-twelfth part, or a 30-degree segment of the wheel, occupies a period of 2,160 years—two more key numbers. If we double these we get two more: 60 degrees and 4,320 years—and so on.

The fact that these same numbers occur in so many diverse myths and legends not only indicates that ancient peoples were aware of precession, but also that the myths themselves very probably have a common origin. Such beginnings, however, reach back to a time too remote for us to identify precisely. Perhaps the myths originated with the legendary Egyptian harbingers of wisdom, Osiris and his grand vizier, Thoth. Certainly precessional numbers appear in many of the myths surrounding these "gods" of the First Time. Then again, possibly this knowledge had its source with the race of ancient mariners responsible for mapping Antarctica way back when the continent was free of ice. As experienced navigators, these people would presumably have developed an acute awareness of the gradually changing star patterns of the night sky. Perhaps Osiris and his companions were actually connected with, or were either descendants or ancestors of, this prehistoric brotherhood of map-making mathematicians and geometers. Alternatively it may be that the true origin of the knowledge of precession dates back to a time more remote than anyone has hitherto imagined.

Certainly observation of the sun, moon, and stars is one of man's oldest pastimes. We see evidence of this in the alignments of many ancient sites, not just in Egypt, but all over the world, in Western Europe's ancient stone circles, the citadels and plains of South America and in the temples and pyramids of present-day Mexico and Guatemala. Furthermore, as Colin Wilson and Rand Flem-Ath have noted in their book *The Atlantis Blueprint,* it is evident that very many of these sites were not simply selected at random, but were chosen to conform to an overall global pattern of longitudinal and latitudinal coordinates. That is, from sacred centers in Egypt and the Americas, to remote Pacific Islands, through ancient Greece and the Middle East and Tibet, there has emerged a clear pattern of whole-number coordinates linking many of them.[6] Obviously these sites are not all contemporary with one another, ranging in age, according to orthodox chronology, from one to five thousand years. However, given that some of these important centers of culture are extremely ancient, it is possible that the original geophysical or metrological plan was mapped out in the very early days, when civilization was supposedly in its infancy.   *10,000 BC ?!*

What I believe is unfolding here, in the light of all the recent research into "prehistory," is a picture of an ancient people who, whatever the precise age they might have lived in, were almost totally preoccupied with the idea of harmony and order. As we have noted, this way of seeing the world reached its peak in the civilization of the ancient Egyptians, who were so evidently concerned with the order and movement of the cosmos. They were also, according to the metrologist Livio Stechini, highly skilled in the measurement of the earth and were able to define the extent of their country in relationship to the latitude and longitude of the planet.[7] (According to Charles Piazzi Smyth, the noted nineteenth-century Astronomer Royal of Scotland, the Great Pyramid stands at the exact center of the largest landmass on earth.) Moreover, Stechini has calculated that the base perimeter of the Great Pyramid—921.453 meters—is exactly equal to half a minute of latitude at the equator, so the perimeter is equal to 1/43,200 of the circumference of the earth. This fact has a twofold significance: first, it demonstrates how remarkably knowledgeable these people were about their planet, and second, given that 4,320 is a precessional number, they may have realized also that there exists an intimate, symmetrical connection between the earth and the zodiacal wheel.

But the rise of the Egyptian civilization of the Old Kingdom, as we have noted, marked a time when people probably already knew about the immensely long cycle of precession. They knew about longitude and latitude, shipbuilding and navigating, and techniques of building with giant blocks of hewn stone that even to the present day have not been adequately explained. Most importantly, they were also familiar with *pi* and, of course, the Hermetic Code, which we now know is a virtual blueprint of the genetic code, the code of life itself.

So once again, if we allow a reasonable period for the natural accumulation of all this sophisticated knowledge, we have a clear indication of an extraordinarily advanced civilization existing in the period historians call prehistory, say, between 10,000 and 4000 BCE, or possibly long before.

How long exactly?

As I said previously, it is not my intention here to try to prove that civilized hominid culture is ten or even one hundred thousand years old. In trying to ascertain where the Hermetic Code originated, however, we must delve very deeply into our past: there is, in fact, datable archaeological evidence to suggest that the most fundamental component of the Hermetic Code—the sacred constant, or the unit of the octave—was "sacred" even in the time of Neanderthal man seventy-five thousand years ago.

In his book *Cities of Dreams* (1989), a highly original study of Neanderthal culture, the psychologist and philosopher Stan Gooch describes a remarkable find at Drachenloch in the Swiss Alps, a known Neanderthal bear-hunter site: inside a cave an altar was discovered. It consisted of a rectangular stone-built chest capped with a great stone slab in which had been placed seven bear skulls with their muzzles pointing toward the entrance of the cave. Six more skulls were discovered set in niches in the cave wall behind the altar.

Gooch sees this as a clear indication that the numbers 7 and 13 (7 plus 6) were sacred to the Neanderthal. He notes also that the constellation of the Great Bear contains seven stars, a fact that prompts him to make this rather bold and startling statement: "We can hardly doubt that Neanderthal had already given it this name that almost unimaginable time ago. And so, in our own times, it is still called the Great Bear by ourselves, by the ancient Greeks, by the Romans, the

Hindus, the Ainu, the North American Indians, tribes in Africa, and many others besides."[8]

If Gooch is correct in assuming that this 75,000-year-old altar was intentionally associated with a particular constellation in the sky—that of the Great Bear—then we already have here the beginnings of the science of astronomy. Scholars may argue that there is no real proof that these seven bear skulls, carefully placed in this ceremonial stone chest at a time when the last great ice age was literally raging full-blast outside, were associated in any way with the seven stars of the Great Bear. Surely the cavemen responsible for this irritating anomaly were just adorning their lair and toying unwittingly with a random number—in this case 7. However, as Gooch convincingly demonstrates in chapter 10 of his book, these people were already extremely interested in the heavens and were, in fact, capable of calculating the periodicity of the planets and the long-term cycles of the moon.[9] Remember also that astronomy is in fact one of the most ancient sciences known, and that a detailed knowledge of the 26,000-year cycle of precession is hinted at in mankind's oldest myths. It may be, therefore, that the practice of observation and identification of prominent stars in the ever-changing firmament reaches back in a continuous line to Neanderthal times. After all, only two precessional cycles ago it was the Neanderthal, not the Cro-Magnon race (modern man), who was the dominant hominid species on earth.

And the number 7? According to Gooch, the Neanderthals actually identified three sacred numbers: 3, 7, and 13. Significantly, two of these—3 and 7—are fundamental components of the Hermetic Code; that is, taken together, they are an elementary expression of the *pi* relationship. With regard to the number 13 it is, perhaps, worth noting that all of the world's traditional musical scales are founded on the pentatonic scale and the so-called circle of fifths. A fifth is an "overtone" and is produced by touching a vibrating string lightly at one-fifth its length. The ancient Chinese found that a series of "perfect fifths" will produce twelve separate notes before the notes begin repeating. Set down in pitch series, these twelve separate notes include all the semitones of our westernized octave. The thirteenth note, seven octaves higher, is the same as the first.

Is this merely coincidence? Of course, it could be. But I am offering an alternative hypothesis, based wholly on hermetic/genetic principles, which is that the emergence of these specific numbers in the cultural practices of our hominid predecessors, numbers that so closely reflect the musical symmetry of DNA and the genetic code, may have been a perfectly natural adaptation acted out almost instinctively by perfectly natural people.

It should now be clear, from the evidence discussed so far, that the story of our origins is by no means clear-cut. There are simply too many anomalies in the prevailing picture of events, awkward "facts" that consistently fly in the face of the orthodox view of the evolution of civilization.

We have, for example, the "impossible" medieval maps—portolans—copies of copies which were arguably made many thousands of years ago, possibly before 4000 BCE, by people with a workable knowledge of geometry and trigonometry.

Then there is the weathering of the Sphinx and its related structures, quite different from the exterior erosion patterns of the Fourth Dynasty pyramids and tombs, which suggests a greater age for the Sphinx than historians will currently permit. Robert Schoch's proposal, supported by other geologists, is that the erosion has been caused by heavy rainfall. But very little rain has fallen on Egypt in the last nine thousand years—hence the desert we see today. This obviously raises serious questions in respect of the present dating of the Sphinx and its enclosure, which, archaeologists tell us, was created in the Fourth Dynasty, at a time when Egypt had for many thousands of years been engulfed by desert sands.

The incomparable size of the stone blocks used in some of Egypt's most ancient structures is also hard to explain. There are no intermediaries in terms of size, no other earlier structures of any significance whose form might indicate some kind of gradual, evolutionary development leading to the use of such incredibly huge stone blocks. Another puzzle is the remarkably accurate metrological data encoded in the dimensions of the Great Pyramid, data that indicates that its designers possessed an intimate knowledge of the dimensions of the earth. The fact that Giza, along with dozens of other sacred sites scattered worldwide, is positioned on a giant grid of whole-number latitudinal coordinates, suggests that this knowledge was not exclusive to the designers of the Great Pyramid.

Further, the very detailed and extremely plausible astro-archaeological evidence highlighted by Bauval and Hancock, among others, points to an era in time for the carving of the Sphinx that is too remote for historians even to consider. And yet the evidence speaks for itself. Giza displays so many astro-archaeological features that it is hard to believe they might all be purely coincidental. Further, the inclusion of precessional numbers in ancient myths and legends from all over the world is a clear indication that the science of studying and mapping the heavens was fully developed at a time when the inhabitants of pre-dynastic Egypt supposedly hadn't even begun to domesticate plants and animals.

Finally, and in my view most importantly, we have the evidence of the Hermetic Code, an encoded expression of the two fundamental laws of creation that, sometime around 2500 BCE, appears to have suddenly flowered into a complete, highly articulated belief system. On the other hand, if we accept that the numbers 3, 7 and 13, sacred to the star-gazing Neanderthal, taken together constitute an elementary expression of the *pi* symmetry, we can envisage a long and winding trail of intuitive and scientific discovery stretching back at least seventy-five thousand years.

By now the reader might appreciate how remarkably talented were these early ancestors of ours. Of course, strictly speaking, this is the "alternative" view. Egyptologists, by and large, officially accept none of the theories I have mentioned here. Indeed, many of them still maintain that the Great Pyramid was built solely as a tomb for the megalomaniac King Khufu.

In spite of this intransigent orthodox opposition, however, the alternative concept of an extremely ancient lost civilization is fast gaining ground. This has resulted, not surprisingly, perhaps, in a renewed interest in the writings of the Greek philosopher Plato, and in particular his story of the great catastrophe that destroyed a high civilization known as Atlantis in 9600 BCE. As I write, I know of three forthcoming books by investigative authors that will be dealing with this enduring historical enigma in some detail. Doubtless they will be met with the usual scholarly objections, but these detailed investigations, which have been briefly outlined by the authors concerned in recent lectures in the UK, will certainly pose more challenges to the standard view of history.

Just on the basis of the current evidence, there is a scenario suggesting that the advanced knowledge we usually see as marking the beginning of recorded history by no means represents the dawning of scientific activity, but rather the last vestiges of the science of a prior great and hitherto unidentified culture.

Plato's dialogue on Atlantis, though, is officially just a story, a "fairy tale" concocted by this overworked Greek sage as a form of relaxation, a means of escaping the rigors of disciplined academic life. That may be the case, and the existence or otherwise of Atlantis is not here my primary concern. It is enough to know that the ancient Egyptians of the Fourth Dynasty were scientists of the first order. After all it was here, on the banks of the ancient Nile, that the numerous branches of the knowledge acquired in "prehistory" were subsequently drawn together in a vast intellectual exercise. And what a truly awe-inspiring enterprise this was, combining precessional astronomy, geodetics and metrology, surveying and architecture, a complete understanding of the scientific laws of creation and, incredibly, a method of applying these laws as a way of being, a "religion." It is difficult to imagine how much more in tune with the world and with nature anyone could possibly be. These people were in tune with virtually everything, with the patterns in the skies, the symmetries on the earth, and, most importantly, the rhythm of life itself.

It seems to me, therefore, that one of the most rewarding lines of inquiry is to try to find out what made these great megalithic builders tick.

Let's investigate.

2

A Different Way of Seeing

O f the many recently published books investigating our cultural origins, there is one that is of particular interest with respect to the ideas being investigated here. This is Colin Wilson's *From Atlantis to the Sphinx* (1996), described by its publishers as an attempt to understand how the long-forgotten race of mariners and builders of prehistory "thought, felt, and communicated with the universe." This position does, of course, presuppose that there was once a race of people existing in the so-called Neolithic era who were capable of contemplating the mysteries of nature and the universe.

Wilson begins with the ideas of Schwaller de Lubicz, which are detailed in a book called *Al-Kemi,* written by an American artist called André VandenBroeck, a former friend and pupil of Schwaller. VandenBroeck says that Schwaller believed the ancient Egyptians and their predecessors had a knowledge system that would be unrecognizable by modern man, a different way of looking at things that gave them a unified perspective on the universe and human existence. This ancient system of knowledge, according to Schwaller, provided a "method of accelerating the pace of evolution."[1]

No details of this method are given, presumably because Schwaller didn't have any. But it nevertheless struck a resonant chord in Wilson's

Ref

mind, because the possibility of speeding up the evolutionary processes, particularly those involved in the structure and development of consciousness, has, as he says, been the underlying theme of all his own work.

Significantly, Wilson has also written several pieces on the life and works of Gurdjieff, notably in his highly acclaimed first book *The Outsider* (1957), and in his encyclopedic classic *The Occult* (1973). In a later book, *The War Against Sleep* (1980), he describes Gurdjieff's "system" as "probably the greatest single-handed attempt in the history of human thought to make us aware of the potential of human consciousness."[2] The "sleep" referred to in the title is what both Wilson and Gurdjieff would describe as normal consciousness. This is the kind that sees most of us through every normal day, a kind of low-resonance state of awareness that, at its lowest level, keeps us from bumping into furniture, jumping red lights, or murdering one another wholesale, and, at its peak, enables us to rationalize, to be logical, and so on. And, of course, it doesn't always work.

But what Gurdjieff, Schwaller, and Wilson all say is that there are other, higher states of consciousness that can be reached, and which were somehow attained by such as the ancient Egyptians. Wilson says as much in his introduction to *From Atlantis to the Sphinx:* that in his view the Egyptians understood "some secret of cosmic harmony and its precise vibrations, which enabled them to feel an integral part of the world and nature."[3]

Drawing on an idea first presented by Robert Graves in his book *The White Goddess,* Wilson suggests that there are two fundamental kinds of knowledge—what Graves referred to as solar and lunar. Our modern type of knowledge, he says, is rational, solar, and works with words and concepts, fragmenting and dissecting everything by analysis. By contrast, the knowledge system of ancient civilizations Graves saw as lunar, or a form of perception based on intuition, which somehow grasped things as a whole.

As an illustration of this latter form of perception, Wilson quotes a passage from Ouspensky's book *In Search of the Miraculous,* in which Gurdjieff explains to his pupils what he sees as the distinction between "real art" and "subjective art." According to Gurdjieff, a subjective work of art is merely a random, arbitrary creation, usually conveying very different impressions to different people. Real art, on the other hand, is as objective as any systematic science and invariably creates the same impression in everyone who understands the basic principles of objective expression, or art imbued with real meaning. Gurdjieff said the Sphinx was an example of real art, and that he had seen many others on his wanderings across Asia. No doubt the Great Pyramid would also have been included among these examples, although, rather curiously, he makes little mention of it in his own writings. The point is, both the Sphinx and the Great Pyramid do, in fact, arouse the very same basic response in everyone, persisting into our times in the form of an acute sensation of awe and wonder.

In his lesson, Gurdjieff mentions one particular and rather mysterious work of "objective art," a certain strange statue he and his fellow travellers encountered in the desert at the foot of the Hindu Kush mountains in Afghanistan. At first, they all thought it was simply an ancient depiction of a god or devil. But after a while he and his companions began to feel that this was no ordinary figure, and that its composition was in fact extremely intricate and revealing in its design and structure. Gurdjieff said that they gradually came to realize that there was in fact a complex system of cosmology embodied in this figure, in its legs, in its arms, in its head—everywhere. In the whole statue, he said, there was nothing accidental, no feature without meaning.

Subsequently this sudden awareness of the statue's esoteric content seemed to induce in his group a different and unexpected kind of perception, through which, he says, they were not only able to understand the symbolism of the figure itself, but also, in some strange, "holistic" way, to feel the thoughts and emotions of the people who had created it thousands of years ago.

Gurdjieff was, in fact, an inveterate storyteller (a trait he seems to have inherited from his father, who was an *ashok*—a bard of some renown), and he may well have invented an imaginary statue here purely for the purpose of exposition. But what is important is what is being conveyed in this story: the idea that objective art is based on intuitive, "lunar" knowledge, and unlike ordinary art, presents the viewer with a complete and coherent picture of the content and meaning of the work.

Much of Wilson's book, like Hancock's *Fingerprints of the Gods,* is concerned with the great enigmas of the past, some of which we discussed briefly in the last chapter. But, in the final section, after a long and highly informative journey through ancient history and remote prehistory, he returns to the question that lies at the root of all of his writings: What is consciousness?

Wilson sees the two fundamental kinds of knowledge, solar and lunar, or rational and intuitive, as operating in different regions of the brain, which consists of two major hemispheres: right and left. The right hemisphere, our lunar side, is responsible for our intuitive processes; the left controls our rational thoughts, our modern, solar functions.

As an example of ancient man's "right-brain" consciousness, which, he suggests, still exists in an attenuated form today in certain so-called primitive cultures, Wilson cites observations made by the American anthropologist Edward T. Hall in his book *The Dance of Life* (1983). Hall spent several years studying the religious and social customs of several Native American tribes, in particular the Quiché, who are direct descendants of the ancient Maya, and the Hopi and Pueblo. He discovered that to many of these peoples, time, as we know it, has no meaning. In fact, the Hopi have no word for it and, in their language, verbs have no tenses. They have no yesterday and no tomorrow, perceiving only an "eternal present," in which time virtually stands still.

On reading this, I was immediately struck by this notion of an endless moment, because it sounded more than vaguely familiar. It suggests, in fact, that the shaman of the Hopi, like those of many other

American tribes, are given to using some kind of hallucinogenic agent in certain of their ceremonies. Certainly their view of time—or the absence of it—is very reminiscent of the extratemporal impressions I had during my "experiments" in the sixties and seventies. It also brings to mind a point I made in the Introduction: that the religious and mystical writings of every major culture, from Egypt right through to Islam, all refer repeatedly to the timeless dimension of heaven.

This notion of timelessness is thus an important link between the psychedelic or shamanistic experience and the mystical revelation, and could provide us with a valuable insight into the true nature of right-brain, lunar consciousness.

So what does it mean to experience only an "eternal present"? Can such a reality be defined in a way that even "left-brainers" can comprehend? I believe it can. In fact, as we shall see, modern science has already provided us with a mathematically verifiable model of such a definition.

Obviously, by its very nature, a "timeless" reality is difficult to rationalize, and so might easily be disregarded as the stuff of primitive imagination or temporary hallucinatory madness—too nebulous to be real. But then there is another, equally nebulous, manifestation of human consciousness that is accepted by practically everyone, and this is our intuitive capacity. Intuition is not a logical process, but everyone "knows" it exists. Indeed, science itself thrives on intuition; it has been responsible for some of the most important discoveries of the modern age. And so, today, as particle physicists probe deeper and deeper into the apparently illogical nature of matter, they are only too aware that "intuition" is one of the most effective tools they have.

This is an odd state of affairs, in that we are positing here a thought process shared, though perhaps to greatly varying degrees, by very different psychological types: by the physicist and the shaman, or by what Wilson calls the scientist, the modern thinker, and the artist, the high priest of ancient days. Obviously there is a little bit of the "lost" artist in every scientist; and, as the physicist's probings become ever more surreal and intuitive, the artist's presence grows in stature. Perhaps, then, this is evolution, the process whereby the artist and the scientist come to coexist in equal measures.

According to Gurdjieff and Schwaller, however, ancient man had far more extensive mental powers than we have today, which suggests that there has been more than a slight hiccup somewhere along the way, something that caused an involutionary trend in our development. If true, one wonders how this could have come about, how a people so highly evolved could suddenly just lose the initiative, fail to pass on their knowledge to their descendants, and all but disappear from mankind's race-memory.

One possible reason, as Hancock has suggested, is that there was some great catastrophe, possibly caused by the extreme conditions of the melt-down at the end of the last ice age, that left only a few survivors of the evolved race—an elite, highly resilient minority, castaways in a new and forbidding wilderness populated by fierce and fearful hunter-gatherers. It would have been virtually impossible for these survivors to have immediately passed on their vast wealth of knowledge to primitive tribes. Yet they realized that somehow, if the less fortunate peoples of this earth were to have any chance of evolving at anything other than a snail's pace, this knowledge had to be kept alive. And this meant dealing with, and controlling, the majority population, whose manpower they needed to utilize in their concerted effort to build on a scale so vast that only another great cataclysm could wipe out all traces of their endeavors.

The myths of the Fourth Dynasty Egyptians and the ancient Native Americans both refer to these superior-minded survivors as gods possessed of supernatural powers. These were the gods of the First Time—the "golden age"—who planted the seeds of ultimate wisdom in the minds and hearts of our ancestors, the primitive natives. They created complex cosmological creation myths and subsequently disseminated them worldwide; they built great works of "objective art," monuments with specific dimensions, orientations, and alignments that, once decoded, would reveal every aspect of their knowledge in precise and graphic detail. And this knowledge, wholly encapsulated in the Hermetic Code, was subsequently embodied in the measurements of pyramids and other megalithic structures all over the world, and in virtually every major religious scripture.

And so then they waited, these gods, for the population to evolve. Gods living in eternity can do that. Presumably they are still waiting, waiting for their seminal message to germinate and come to flower. And who knows? Maybe the current upsurge in awareness of the extraordinarily advanced ideas of our most ancient predecessors is the first sign of a new bloom.

Of course, this is just one historical scenario, and it doesn't bring us any nearer to answering the question posed by Wilson, i.e., how might these gods have perceived the world?

Strange as it may seem, we may possibly find at least part of the answer not at the dawn of history, but at the very frontiers of modern science. We have already noted how scientists are becoming, as Graves might say, more and more "lunar" in their mode of thought. The modern scientist is not a god, but he has at least learned to use his intuition in his quest for the truth about reality. And what he has discovered, for example in the microcosmic world of the subatomic particle, is extremely interesting, because the reality now being described in scientific terms brings us full circle, right back to the visions of the Hopi, the hippie, and the Egyptian high priest.

"Down there," in the world of the elementary particle, time has no quantifiable meaning, no value; it simply doesn't exist. This is the real world we are talking about here, the world defined by physicists in precise mathematical terms. Therefore the Hopi, in a very real sense, have it exactly right: timelessness is the primary reality. More than that, the ancient Egyptians, I believe, realized this also, as is evidenced by this ancient poem referring to the god-king:

> *His lifetime is eternity,*
> *the borders of his powers are infinity.*[4]

Lofty thoughts indeed for a people whose ancestors of only five hundred years before were simple, wandering nomads. I shall have more to say on this "modern" notion of timelessness later.

But in fact, as we shall see, there are other crucial shamanistic concepts that are equally at home in the modern scientific mind. Consider this. The most important part of Hopi ceremonial life is the dance—hence the title of Hall's book. He writes that if the dance is performed correctly, to the participants everything—the entire universe—"collapses, and is contained in this one event." Thus the Hopi's experience of this alternative reality is not only timeless, it is spaceless, and implies a dimension in which everything—time, space, matter—can collapse into a single conceptual "eternal moment." Once again we have a close Egyptian parallel in the reference to the god-king quoted above, whose life span we noted was described as eternity, and the borders of whose powers are infinity. Endless time . . . infinite space. . . . And this is just for starters. Later, when we take a more detailed look at other scientific discoveries, we shall see how this "primitive" notion of spacelessness also has quite distinct echoes in the present day.

Many Native American tribes consider the earth to be a living, matriarchal being. Some believe that she becomes pregnant every spring and should therefore be treated gently. Thus they will remove steel shoes from their horses and modern shoes from their own feet for fear of breaking the surface of the earth. As Wilson says, such a notion is not simply an idea or belief, but "something they feel in their bones, so that an Indian's relationship with the Earth is as intimate as his relationship with his horse. . . . To regard this as a 'belief' is to miss a whole dimension of reality."[5]

Wilson believes that the ancient Egyptians shared a similar kind of intimacy with nature, and with the land and skies of the Nile valley, perhaps partly due to their close relationship with the life-giving River Nile, whose annual inundation occurred just as the star Sirius/Sothis returned to the early morning sky after seventy days spent below the horizon. These regular and important events would have ensured that the Egyptians remained very much in tune with their environment, with the earth, with the sky, and with the rhythmic unfolding of the seasons.

So what Wilson is saying is that Egyptian knowledge was not simply based on superstition, but on "a deeply experienced relationship with the earth and the heavens."[6] As with the Native Americans, this contact with the world about them was something the Egyptians felt in their bones, and what they felt was the rhythm of nature. Schwaller shared a similar view, stating that "every living being is in contact with all the rhythms and harmonies of all the energies in the universe."[7] But Schwaller also believed that modern man had lost touch with nature's rhythms and harmonies, and when he speaks of ancient knowledge providing a method of speeding up the evolutionary process, he clearly associates this method with the reestablishment of man's former, intimate relationship with nature.

Now, Schwaller's ideas are sound enough to deserve consideration, but Wilson, while obviously sharing similar views, was prompted to pose the following question: "But is there any way to turn this rather vague and abstract statement into something more concrete and down to earth?"[8]

In my view, there is, a wholly practical way of looking at the processes of the development of consciousness, the very "secret of cosmic harmony and its precise vibrations" that Wilson believes the Egyptians possessed. The Hermetic Code, the musical theory of transcendental evolution embodied in Egyptian myth and religion, in the I Ching, and in just about every major religious doctrine known, fits the bill perfectly. Wilson doesn't actually say as much, but he then goes on to devote several pages to an impartial commentary on my first book, including details of the close correspondences between the structure of the I Ching, the *pi* symmetry, and DNA and the genetic code. I have to say that his studied appraisal of the ideas presented in *The Infinite Harmony* is very gratifying—particularly so since, as far as I know, he is the only published writer to have broached them to date. Whether he agrees with everything I say is, of course, another matter. When I spoke with him prior to publication of his book he hinted then that there would be no committal, no outright endorsement of my ideas. Typically, he has been true to his word. Nevertheless, this now gives me an opportunity to strengthen my case by filling in some of the gaps left in his commentary.

As we have noted, the musical symmetry described by the Hermetic Code is echoed, note for note, in our very bones, in our DNA and in the genetic code. Through the world's religions, mankind has been instinctively living out the basic principles of this code for thousands of years. This is because it is a perfectly natural thing to do; it is the way of creation. The trouble is that somewhere along the line we forgot why we were holding the seventh day sacred and acting out "passions." We have had timely reminders of our sorry state from individuals like Moses, Christ, the Buddha, Zoroaster, and Muhammad, which have helped to keep "the faith" alive, but such has been our lot that we have tended always to forget and switch back over to automatic pilot. This is how a great work of "living music" like the I Ching can end up being used as little more than a pocket fortune-teller.

The Egyptians, I believe, never forgot, but consciously persisted in the application of the principles of the Hermetic Code as a complete mode of being, a "religion" in the fullest sense. This is how they were able to develop their acute sense of belonging with the world and so become increasingly more conscious in it. I think this is precisely the "method" Schwaller was looking for.

The remarkable thing about the Hermetic Code is that it is not only a blueprint for our inner development; it also has very definite cosmic applications that provide us with a direct link to what Schwaller called "all the rhythms and harmonies" of the universe. This is how.

The theory of transcendental evolution, as I have said, is based on the knowledge of the structure of the major musical scale, and the idea that all life evolves ever upward, as the notes in a developing octave evolve into higher, more resonant scales of existence. Going up, in other words, toward the stars. In a broad sense, the neo-Darwinian theory of evolution accords with this concept. Over billions of years, bacteria have evolved "upward" into single-celled organisms; they in turn have evolved into multicellular structures, and these have evolved into all the complex animal life forms existing today. This continuous chain of evolution, from the free-floating bacterium to the free-thinking human being, is a transcendental process involving a continuous succession of quantum leaps from one scale of existence into a greater scale above.

I'm not suggesting, of course, that the ancient Egyptians knew all about biomolecular events in the microcosm—at least, not through an accumulation of fragmented facts experimentally verified by prehistoric biologists and geneticists. What I am saying, however, is that the Egyptians probably took all this for granted, that they already intuitively understood what was going on inside them, or "down there." As a matter of fact, the originator of the Hermetic Code—Thoth, Hermes, whoever—left us with one simple yet incredibly astute dictum that sums up the theory of transcendental evolution perfectly: "As above, so below."

"Above" we have the Hermetic Code as expressed through the *pi* convention, 22/7, composed of three octaves, each of which is itself composed of three octaves, a total of sixty-four inner notes. "Below" we have the genetic code, whose structure is not merely similar to the Hermetic Code, but is identical right down to the very last detail. Not only that, but both codes, as we noted when discussing the I Ching in the introduction, share a common purpose, which is to facilitate the processes of creation, of evolution onto a higher scale of existence. In the case of the genetic code working within your body's cells, this higher scale is represented by your whole being and, in particular, your mind. But in the case of the Hermetic Code operating within "cells" of some kind in an infinitely greater "body," the next, higher scale must lie far beyond the confines of the brain, somewhere . . . up there.

The established neo-Darwinian theory of evolution, of natural

selection through random mutation, attempts to explain only the evolutionary development of physical bodies in the local biosphere of our planet. It is an unfinished theory. The theory of transcendental evolution, however, gives us the whole picture; it tells us that the evolutionary chain of life doesn't end with the human being struggling to survive in a competitive terrestrial environment. The truth is, where the ongoing evolution of mankind is concerned, as the Egyptians knew full well, the possibilities are limitless. That is, nature's evolutionary processes, encoded within the musical structure of DNA and the genetic code, continue to evolve higher still, beyond the confines of the physical brain encased in the skull, into metaphysical scales of existence that ultimately encompass the entire universe.

We noted in the previous chapter that ancient man was intensely interested in the stars, in precession, and in immensely long cycles of time. Now we can see why. The ancient Egyptians weren't simply admiring the view; they were staking their claim in the greater scale above, paving a way to heaven. Cosmology with a capital C.

In the world of computers, virtual reality, and endless information highways, the modern mind is more often than not inclined to look back on these times with superiority. Even today, historians continue to portray ancient Egyptians as manic "tomb builders," highly gifted but superstitious stargazers, whose monumental architectural designs, however ingenious, are totally devoid of any esoteric meaning. But as we have seen, these remarkable people in fact had a complex cosmology superior even to our own, one that included within it the cosmologist himself.

Our modern cosmologists assiduously study everything "up there." That is their job. What lies "below" is no concern of theirs. But this partial view presents problems, as with the Egyptologist who meticulously searches the sands for telltale shards of pottery and other fragments, but who knows nothing about astronomy. This is the tunnel vision of specialization that prevents many "experts" from seeing holistically. As a result of this imbalance, scientific knowledge has become segregated; it is not a part of our everyday lives. We might look up and think about astronomical questions occasionally, but on the whole our world is predominantly terrestrial, confined and circumscribed in relation to the vastness of the universe around us.

Now consider the world the ancient Egyptians inhabited, which not only included the land of their birth, but also the earth itself, the sun, moon, and planets, and even the ever-changing constellations. This is an all-encompassing vision. And the civilization capable of perceiving it showed its stature through the grandeur of this vision. The Egyptians had their sights set firmly on the heavens, their sole raison d'être being an intense, concerted effort to assist mankind on its transcendental journey to the stars, to the greater scale above. And to make absolutely sure that their message carried, they skillfully encoded the fundamental principles of their extraordinary life science in the dimensions, proportions, and alignments of massive stone monuments built with unsurpassable precision, many still acknowledged as the greatest man-made structures on earth.

As we shall see later, the cosmological perspective of Egyptian metaphysics is, in fact, the basis of ancient man's entire belief system, and may even be, as many commentators now suspect to be the case, a legacy from an even earlier period of civilization existing in what we currently refer to as prehistory. But before we can ascertain how their musical method of self-development might have given them direct access to the greater cosmos, we first need to consider more earthly matters, which will be the subject of the next chapter. The main question arises from the established historical fact that virtually all of the cyclopean monuments created at the dawn of our history were built before the wheel was invented. So how was this accomplished?

3

Music over Matter

No one has yet convincingly explained how the architects and engineers of the ancient world managed to build on such a monumental scale. This applies equally to structures on both sides of the Atlantic, many of which have been shown to display remarkably similar design features. In the ancient city of Tiahuanaco in the Andes, as in Egypt, blocks weighing more than two hundred tons are commonplace. There is one construction block that has been estimated by Graham Hancock to weigh as much as 440 tons. Also in Peru, the citadels of Sacsayhuaman and Machu Picchu contain similar megalithic stones that have been cut and carefully positioned with a degree of precision that even modern construction engineers would be hard pressed to match.

Equally mysterious is the presence of truly giant pyramids on both continents. The Great Pyramid is arguably the most notable, being the largest solid stone edifice ever constructed by man, having a base area of over thirteen acres. However, the great Pyramid of the Maya at Cholula in Mexico, despite its core consisting not of blocks of stone but of rubble, is in fact more than three times as massive as the Great Pyramid. Its base covers an area of forty-five acres, making it easily the largest building on the planet.

Now, fashioning and carefully placing hundreds upon hundreds of

huge blocks of stone to conform to precise geometrical and astronomical alignments is an art in itself, but the immense scale of these enterprises is not the only puzzle. There is also the question of how these
craftsmen actually carved and cut the stone. William Flinders Petrie
examined the red granite "sarcophagus" in the King's Chamber of the
Great Pyramid and noted that it had been hollowed out to such a fine
degree of accuracy that its external volume is exactly twice its internal volume. And this was achieved with one of the hardest stones on
earth. The method used, according to Petrie, was some kind of tube-
drilling mechanism, rather like a section of drainpipe with exceptionally hard teeth set into the rim. How such an implement might have
been powered is a question that only compounds the mystery.

Petrie surmised that the sarcophagus itself had been cut from the
mother block with a "saw" at least two and a half meters long, though
no evidence of such an implement has ever been found—except for the
serrated marks, of course. Petrie also found evidence of the use of circular saws and even lathes; but again, only the manufacturing marks on
numerous stone artifacts remain as proof that such tools ever existed.

In any event, even if drills and lathes were in common use in
ancient Egypt, this would not explain the discovery of a large number
of hollowed-out basalt vases found in and around the Third Dynasty
Saqqarah necropolis and dated to around 4000–3000 BCE. As
Hancock describes in *Fingerprints of the Gods,* some of these elegantly
curved vessels with widely flared interiors have long, slender necks too
narrow for even a small finger to be inserted. And yet they have somehow been hollowed out with unbelievable precision.

It has been suggested by a modern toolmaker called Christopher
Dunn, who has studied the baffling stonework of the Egyptians in
some detail, that the craftsmen responsible for some of the work may
have had a technology based on high-frequency sound. Basically Dunn
believes that the workmen may have employed some kind of ultrasonic
tool bit capable of vibrating at a rate thousands of times faster than a
pneumatic drill. However, even if such a mechanism were used, as in

the case of the proposed drills and saws and lathes mentioned above, we would have to assume that the Egyptians could somehow produce the power necessary to drive such devices.

Whatever may be the case, it is becoming increasingly evident that, contrary to the beliefs of orthodox archaeologists, the craftsmen responsible for some of these mysterious artifacts certainly did not use crude copper chisels, adzes, and simple wooden mallets to do the job. As Dunn has noted, these people were capable of producing smooth, flat surfaces on granite or basalt to an incredible accuracy of a thousandth of an inch or more. He demonstrated this to Robert Bauval in the Cairo Museum by placing a high-precision metal gauge against a side of the ancient relic known as the Ben-Ben Stone and shining a light against the line of contact. That no light was visible from the other side indicates an engineering accuracy equal to that of the present day.

Perhaps one of the most interesting aspects of Dunn's idea of the use of high-frequency sound is that it implies the application of a concentrated form of resonance, the raw material of music, the principles of which, as we have seen, were the mainstay of Egyptian metaphysics. So possibly these people used their knowledge of cosmic harmony and vibrations to assess—or maybe "feel"—what kind of ultrasound frequency would be required to work a given material. After all, everything vibrates, resonates, and it may be that the Egyptians had found a way of creating sympathetic frequencies that depended not so much on a powerful energy supply but more on an understanding of the subtle, interrelated structures of these inherent symmetries.

Such a notion may at first seem somewhat abstract, but later we shall see that the Hermetic Code provides us with a rather ingenious description of the mechanism by which these "inner" symmetries can be understood. This is further supported by the ideas of Gurdjieff, some of which I shall be discussing in detail in chapter 12, and also by some of the ideas of a number of modern scientists in disciplines as diverse as clinical psychology, neurophysiology, and even that seemingly inviolable sanctum of all empirical science, nuclear physics.

Unfortunately Dunn's ultrasound theory, if correct, would still not explain how such massive stones, such as those incorporated in the Valley Temple at Giza and the Osireion at Abydos, once cut and shaped, were then moved into position.

Colin Wilson has suggested that the builders might have employed a method similar to the popular party trick in which a subject sits on a chair and four volunteers place one finger underneath each armpit and knee and try to lift him or her. Without any preparation, the result is as one might expect, and the subject, unaffected, stays put. If, subsequently, the volunteers all place their hands on the top of the subject's head, first their right hands and then their left hands, and then concentrate hard for a minute or so, when they simultaneously remove their hands and try once more, the subject can sometimes be lifted high off the ground with very little apparent effort. It's as if four people concentrating in unison can somehow exert a new, much more powerful kind of force. One volunteer alone would find difficulty in lifting a quarter of the weight of a fully grown subject with a single index finger, yet four together can not only lift four times that weight, they can often do it with astonishing ease.

Wilson suggests that this "group-mind" phenomenon was possibly a basic way of life to the ancient builders of Egypt, who saw nothing extraordinary in moving great chunks of stone in this way, perhaps believing that the gods were making the blocks lighter, and that no special effort was required other than acting in unison, in harmony with one another.

With regard to the party trick mentioned above, one wonders whether the mind of the subject might also be involved. There is a certain amount of pressure on the subject's head from the hands of the volunteers, and when that pressure is released the subject naturally feels a sensation of becoming suddenly lighter. If this were a contributory factor in the experiment, it would raise further questions relating to the raising of inanimate blocks of stone, which could not, one would assume, participate in the experiment in any way.

Extensive tests have, in fact, been conducted under laboratory conditions, the results of which indicate that psychokinesis, or the ability to affect physical objects with the mind, is in fact a statistically verifiable reality.

In a series of experiments conducted in the 1970s, Robert Jahn of the Princeton University School of Engineering and Applied Science and the clinical psychologist Brenda Dunne used an instrument called a random-event generator (REG) to test the psychokinetic abilities of a large number of volunteers. Triggered by the process of decay of a radioactive material, which is an unpredictable, natural process, a REG is an automatic selector—a "coin flipper"—that produces a completely random series of binary numbers. Volunteers were asked to sit in front of the device and concentrate on trying to make it produce an abnormally large number of either "heads" (1) or "tails" (2). Subsequently Jahn and Dunne's results clearly showed that, simply by concentrating on the REG, the volunteers were able to influence the binary output to a small but statistically significant degree.

In another series of tests they used a kind of pinball machine in which 9,000 marbles were allowed to roll around 330 nylon pegs and cascade out of 19 exit holes into bins. Once again, over the course of many trials, they found that most of the subjects were able to produce a small but significant change in the average number of balls falling into each bin.

Jahn, a professor of aerospace sciences, was at first skeptical and reluctant to involve himself with these experiments, but he was eventually so impressed by the results that in 1979 he founded the Princeton Engineering Anomalies Research Laboratory, where researchers have continued to produce strong statistical evidence in favor of the existence of psychokinesis. Furthermore, as the above-mentioned experiments have shown, it looks as though the ability to produce detectable psychokinetic influences is not limited to the few, but is something that most of us can do.

Another, rather more unusual example of psychokinetic abilities is that which physicists call the Pauli effect, where merely the presence of certain individuals appears to cause machinery and equipment to malfunction. The classic example is the eponymous physicist Wolfgang Pauli, who would often and unwittingly cause sensitive equipment to go wrong, or glass apparatus to explode, simply by being there.

This effect may be similar, though not identical, to the one I described in *The Infinite Harmony*, where the psychologist Carl Jung caused a piece of heavy furniture to split apart with a force that appeared to emanate from his solar plexus. Sigmund Freud actually witnessed the incident, but when Jung declared that he had somehow been responsible, Freud bluntly refused to believe it.

There are fundamental differences between this "Jung effect" and that of Pauli. First, Jung was conscious of it as it was happening, because the "force" caused his solar plexus to heat up; second, while Freud was shaking his head in disbelief, Jung, feeling the force build up once more, was able to predict that it would happen again. Within moments there was another loud crack and the wood splintered a second time.[1]

Obviously these examples are not quite the same as the party-trick phenomenon, which relies on the concerted, conscious effort of a group, but they at least serve to remind us that there is probably a great deal more to the power of the human psyche than its recognized potential for reasoning, inventing, conceptualizing, and so forth.

When considering the possibility of the "group-mind" technique, Colin Wilson is not implying that the architects of ancient Egypt actually levitated their megalithic blocks. Rather, he thinks that the cumulative power of "group consciousness" was an everyday reality to the Egyptians, and that apparent physical forces, when applied in concert by a given group in this way (perhaps orchestrated by priests uttering magical incantations), could somehow be magnified to an unusually high degree.

Now it so happens that there is a very close parallel to Wilson's idea in the legends and myths of the ancient Greeks. The Pythagorean philosophers, who clearly inherited their hermetic or "musical" knowledge from

Egyptian sources, had a name for this "group consciousness," *homonoia,* which translates as a "union of minds." This state could be achieved, they believed, by emulating the divine actions of the god Apollo and the nine muses, the patrons of all the arts. Significantly, Apollo himself was depicted as the supreme musician, and the word *music,* derived from the word *muse,* originally referred to all aspects of learning. According to the Greeks, through this kind of harmonious union of minds, mankind could literally change the world.

The Pythagoreans believed that their mythological heroes and gods— including Pythagoras himself—were able to play special forms of music that could directly affect both sentient beings and non-sentient things. Orpheus, for example, is said to have moved rocks and even mountains with the power of his music; and another legendary hero, Amphion, with only his lyre moved rocks and stones to construct the walls of the ancient city of Thebes.

Interestingly there are exact parallels between these and some legends of the peoples of the pre-Columbian Americas. For example, one Mayan legend says that the construction of the Pyramid of the Magician at Uxmal in Yucatán was a relatively simple affair, that all the builders had to do was "whistle and the heavy blocks would move into place."[2] Another tradition tells how the blocks used in the building of the city of Tiahuanaco in the Andes were "carried through the air to the sound of a trumpet."[3]

In his book *In Search of the Miraculous,* P. D. Ouspensky recounts a talk given by Gurdjieff concerning the literal truth of these musical myths. Gurdjieff states that "objective music," as he calls it, cannot only destroy, as in the case of Jericho, but also create. Orpheus disseminating knowledge simply by playing the lyre is cited as an example, and further, he says, there could be such music as would freeze water, or even kill a man instantaneously.

Though not quite so dramatic, we see examples today of music, or sound vibrations, having a direct effect on people and things. Most people have heard of the soprano's voice that, when pitched at a given level, can shatter a glass. Of course, a glass is an inanimate object, but certain sounds can also have a dramatic effect on people.

In particular, a new innovation in surgery known as high-intensity focused ultrasound (HIFU) involves using powerful beams of ultrasound as a "virtual scalpel" that allows surgeons to operate deep inside the human body and to target very accurately and destroy malignant tumors. The precision of this technique is such that, unlike established cancer treatments like radiotherapy and chemotherapy, the operation can be carried out without damaging any of the surrounding tissue. And there are no apparent side effects. Early trials have proved very promising. To date, almost one hundred people in Britain have been treated with ultrasound for liver cancer, with positive preliminary results.[4]

Another example is the kind of sound that has recently been employed by a Swedish manufacturer of burglar alarms. It resembles that excruciating sound of a nail scraping down a blackboard, but not just one nail, many of them, with the sound magnified to an intolerable degree. Subjects in trials invariably ran out of the sound room within seconds, while those who tried to stick it out reported feeling physically sick. The maximum time anyone cared to endure it for was around fifty seconds. Personally I have not heard the sound produced by this diabolical machine, but I have been around enough blackboards to know that I certainly don't want to. I only have to sit here and imagine it and I can literally make my teeth go on edge. This kind of "music" might not directly kill you, but, if prolonged, it could very possibly drive you mad enough to do the deed yourself.

In one of his own books, *Beelzebub's Tales,* in chapter 41, "The Bokharian Dervish," Gurdjieff speaks about a demonstration of "objective music" given to him by cave-dwelling ascetics in the mountains of Central Asia. During the course of the experiment, which involved playing obscure sequences of notes on an elaborately modified piano, he watched as a large abscess rapidly appeared on the leg of one of those present. When another series of notes was subsequently played, the abscess, which evidently caused the subject very real pain and discomfort, mysteriously faded away. In another demonstration, fresh flowers were made to wither and die within minutes.

Of course, the ancient musical legends describing the extraordinary abilities of master stonemasons could be pure fiction, although it is difficult to understand why so many identical myths—and there are many—should have emerged in such widely separated regions. And then we have the amazing architectural evidence itself, in sites the world over, which demonstrates a superior and currently inexplicable technological proficiency in the handling of stone.

All this actually proves nothing, however, and those of a scientific turn of mind will feel that such tales of "musical magic" are either allegorical, or that they are simply the stuff of imagination and superstition. Perhaps so. But even those at the cutting edge of scientific enquiry would agree that all their sacred laws are not yet written. There is currently in circulation a whole new batch of fantastic ideas concerning the nature of universal reality. These range from the macroworld of chaos and complexity theories implying an underlying cosmic unity, to the microworlds of superstrings and inter-penetrating loops in which even so-called empty space is seen as a woven fabric of unimaginably fine threads of . . . well, something or other. As we shall see in subsequent chapters, there is no end to the ingenuity and imagination of scientists intent on discovering the "theory of everything."

Given such an open-ended scientific view of the world, it would seem that there is still time and space enough to accommodate the ancient notion of "musical magic." After all, this "primitive superstition" has already found expression in the field of biochemistry, where we see that the respective musical symmetries of the genetic code and the Hermetic Code are identical in every respect. So then we hear that the people who first revealed the Hermetic Code also believed in the power of a strange kind of music that could "enchant" just about everything: trees, wild beasts, even rocks of the hardest stone. We can see how this could apply to trees and wild beasts, because the very essence of both the animal and vegetable kingdoms is music; it is the genetic code, which "enchants" just about every living thing. Rocks are another matter; they are not imbued with life as we know it. Could it be, then, that they are imbued with some form of life as we don't know it? This is what the mythmakers say.

The Greek philosopher Thales, reputedly one of the teachers of Pythagoras, taught that the whole universe was alive and that even inanimate things like rocks and mountains possessed psychic attributes. The "builder gods," Orpheus and Amphion, were said to have had the ability to tune in to this elemental consciousness, and so persuade inanimate objects to do their bidding. This notion could easily be discarded as simply another example of primitive superstition, until we learn that the idea that elementary atoms of matter might possess some kind of awareness of the world about them is now being seriously considered by physicists. Later we shall be exploring this newly discovered "quantum" reality in more detail. As we shall see, many of the scientific discoveries relating to this microcosmic wonderland may not, in fact, be quite as new and radical as most scientists believe.

Thus, according to Thales, the ancient builders of Egypt did not see things in the same way as we do. To them, everything was to some degree alive, conscious, "psychic." Interestingly, this is an idea that is reflected quite clearly in the long-standing traditions of the great Indian yogi masters, many of whom are reputed to have possessed psychokinetic and telepathic powers.

The eminent Indian philosopher Sri Aurobindo, echoing Thales, stated that psychokinesis is possible only because matter is to some extent conscious. If matter were truly inert and lifeless, there would be no conceivable means of contact between the thinker and the object. Further, if a single point in the universe possessed zero consciousness, he said, then the whole universe itself would have to be unconscious.

Similarly, in his book *Autobiography of a Yogi*, Paramhansa Yogananda describes meeting numerous yogi masters who could materialize and dematerialize their own bodies and other objects at will. He claimed that such holy men can actually move at the speed of light and utilize "creative light rays" to bring into instant visibility any physical manifestation.[5]

Obviously Yogananda was referring here to a more subtle form of materiality than sandstone or granite, and this is a difficult concept for our logical minds to accept. But the world of the Indian yogi, like that of the Hopi shaman, is primarily a world of the mind, an alternative reality in which psychokinesis is seen as a perfectly accessible human function. This kind of abstract notion might seem far removed from the question of building with megalithic blocks of stone, but there is, nevertheless, an extremely important link between Yogananda's view of the phenomenon of "creative" light and that of the ancient Egyptians, a link fundamental to our understanding of the whole cosmology of ancient man. We shall shortly be looking at this crucial connection in some detail.

Now, I'm not suggesting for a moment that buildings like the Great Pyramid and the Sphinx enclosure were simply "thought up" by gods, at least not without human intervention. But from the evidence at hand I feel reluctant to accept the uncompromising view of orthodox archaeologists who insist that there was nothing unusual about the methods of construction employed. Many of the stone blocks used by these ancient builders, remember, are several hundred tons in weight. The largest so far identified is the massive, free-standing foundation stone of the Temple of Baalbek in Lebanon, which is estimated to weigh a staggering 1,200 tons. Cutting and shaping a megalith of such monumental proportions is in itself an accomplishment that makes the patchwork concrete and steel structures of modern builders look positively Lilliputian, but then to move this block hundreds of meters to its present location is a technical maneuver that practically defies belief. The most powerful lifting gear used in the modern construction industry can lift hundreds of tons, and a trained team of workers have to spend weeks preparing the ground of the proposed site beforehand, examining the subsurface, leveling, laying down hard core, and so forth. Yet here we have a single, free-standing megalith whose estimated mass is over twice the maximum lifting capacity of even the most modern boom-crane. This in itself does not, of course, constitute proof of the use of paranormal powers on the part of the people who executed this remarkable feat, but it nevertheless raises fundamental questions concerning the orthodox view, that this block was moved purely by conventional means.

Colin Wilson, as we noted, has suggested that the "group consciousness" technique might have been employed in the manipulation of such blocks. He believes that the builders still used conventional means of construction—ramps, rollers, levers, ropes, and the like—but that the real power behind these methods emanated from the collective "vibrations" of the "group mind." And this is precisely what is being implied by the Pythagoreans' notion of a state of homonoia. To these thinkers, as with their Egyptian forerunners, homonoia, a collective unification of mind, body, and spirit, was an attainable reality. They believed that, when totally synchronized—as it apparently was in the old mystery schools of Orphic origin—the psychic energy generated in this way could somehow be used directly to influence matter, even gargantuan lumps of hard rock.

As Jahn and Dunne have demonstrated through their extensive tests, statistically discernible psychokinetic powers can be exhibited today by perfectly normal subjects using nothing more than their ordinary concentration. In other instances, such as the Pauli and Jung effects mentioned earlier, the results are sometimes dramatic, but often erratic and uncontrolled. Nevertheless, there are thousands of accounts from every age describing the paranormal powers of gifted individuals: in the legends of the builders of antiquity, in the scriptures of virtually every major religion, and in the many stories of the lives of saints and "psychics" the world over. Even supposing that many of these accounts might have been invented for effect or whatever, one feels that it would be stretching credulity too far to presume that there wasn't a single grain of truth anywhere in the vast store of literature on the subject.

In the party-trick phenomenon, in which a set of minds enter into a homonoic state, it seems as if the cumulative force so generated is considerably greater than the sum of its parts. The volunteers being raised in this kind of experiment can often be lifted high off the ground with such disproportionate ease that they seem almost to be floating. And if this same method could also be used effectively on inanimate objects, then possibly, given the right circumstances, a large mass would require relatively few people to shift it. This is interesting, because it may provide a possible answer to a puzzling question raised by John Anthony West in a 1992 documentary film about Robert Schoch's investigations into the weathering of the Sphinx enclosure. West noted that the enclosure, which is surrounded by steep-sided natural bedrock, is relatively small in view of the enormous size of the stone blocks that had to be maneuvered within it. There would have been insufficient room for teams of workers large enough even to drag a two-hundred-ton block, let alone lift it. If, however, something like the "group-mind" technique were the force behind their ropes and levers, a force considerably greater than the sum of its parts, then the paucity of working space might not have been a problem.

In any event, we have already ruled out the use of massive lifting devices and excessively large numbers of manual workers, so there has to be some other explanation: the "group-mind" technique, which, as we know, works on living subjects, is at least a theoretical possibility. And, as we have noted, something very similar was involved in the early Greek concept of a joint state of homonoia through which, it was believed, mankind could ultimately transcend to greater things, some kind of collective psychological harmony acquired via a thought system based, according to the Greeks, on the music played by the god Apollo. And this "music," the art of the muses, was not simply concerned with the theoretical aspects of the science of harmonics—that is, the systematic definition by the Pythagoreans of the mathematical structure of the major musical scale—but also with an awareness of the greater cosmic order, with knowledge of the principles and practical applications of the Hermetic Code. It is this concept, I believe, that in some mysterious way lies at the root of the special form of music played by the heroes, Orpheus and Amphion, by the stonemasons of ancient Egypt, and by the "builder gods" of Central and South America.

So did these people really possess supernatural powers? We may possibly never know, but we are certain that they possessed extraordinary abilities. By the uninitiated, these highly advanced skills could easily have been construed as magic. Of course, the exact methods of construction used in ancient Egypt elude us still, and we ourselves, for all our accumulation of technological expertise, are left in a position not dissimilar to that of the early propagators of ancient myth, who evidently witnessed the actions of this highly developed people, but did not fully comprehend what they saw. Modern observers, who do not, by and large, believe in magic, obliquely refer to this forgotten science as the use of "unknown techniques."

We noted earlier that all around the world there are legends and myths speaking of a time long ago when godlike civilizers used the power of music to build the first cities. According to such stories, these mysterious builders could move great blocks of stone simply by creating special forms of sound, by playing musical instruments, whistling, singing, or whatever. In Mexico, Bolivia, and Peru, and in numerous regions in Central Asia, where these legends abound, there is a single common factor that gives credence to all of them: the hard evidence, in all of these locations, of buildings incorporating truly gigantic stone blocks.

In his book *Gods of Eden,* Andrew Collins devotes three chapters to the subject of what he calls sonic technology, citing some of the mythological accounts already quoted, but including also accounts of travelers to Tibet in the first half of the twentieth century who witnessed the apparent levitation of stone blocks actuated by monks using numerous, specially contrived sound instruments. Published in the 1950s by a Swedish engineer, Henry Kjellson, one of these accounts concerns another Swede, a certain Dr. Jarl, who was invited by a Tibetan acquaintance to visit him at his monastery near Lhasa."

One day during this visit Jarl accompanied about 240 monks to a nearby meadow adjacent to a high cliff face. About 250 meters up the cliff was an entrance to a large cave, on the outer ledge of which were several other monks. Forty or so of the monks assembled below took up strategic positions in slightly more than a 90-degree arc around a large, cupped, stone platform. They then began to prepare a large number of instruments: thirteen drums, of varying size, with a skin at one end and open at the other, and six "ragdons," described as three-meter-long trumpets. Subsequently a large stone about one and a half meters in length and one meter in height and width was dragged by yak to the cupped stone platform and manhandled onto it by attendant monks. The musicians then began playing, at first slowly and rhythmically, apparently "pointing" their instruments at the stone at the apex of the triangular shaped assembly. Gradually the noise from the drums and trumpets increased and then the tempo sped up so quickly that Jarl lost track of any rhythm. His account of what happened next sounds like pure fantasy. Allegedly the stone in focus at first began to wobble and then it rose from the ground with a rocking motion. As it rose, the drums and trumpets were tilted upward, aimed constantly toward the stone, which continued to rise in a long parabolic arc, until it ultimately crashed down with considerable force onto the ledge at the mouth of the cave, 250 meters up the nearby cliff face. For much of that day Jarl watched as the process was repeated five or six times an hour.[6]

As if. This is the response I would expect from most of you. Indeed, on first reading this account I experienced the same old knee-jerk reaction myself, living as I do in a predominantly secular environment, where "miracles" such as the one just described occur only in fairy tales. But at the back of my mind, I have this confounding piece of evidence, an undeniably real artifact, whose very existence gnaws at the core of my reason. This is the 1200-ton stone megalith of the Temple of Baalbek in Lebanon, a perfectly shaped block almost twenty-five meters in length, with a mass nearly two and a half times greater than anything that could be lifted by the largest boom-crane on Earth. At some period in ancient history, long predating the Greek structures on the site and before even the simple wheel had been invented, this veritable monster was somehow transported several hundred meters from the quarry of its origin to its present location high above sea level. Three comparatively smaller stones, each weighing something approaching 600 tons, were also carved and transported with it.[7]

So we have hard evidence, great, monstrous lumps of solid stone standing as high as five-story houses, which attests to a stone-raising technology vastly more sophisticated than our own. We might expect that modern engineers could, given sufficient time and funds, build a boom-crane capable of raising such a mass upward, but by what conceivable means could they then, without the application of the wheel, apply a sideways motion to these megaliths, covering not just a few yards, but hundreds of meters of undulating terrain? It seems to me that until the experts can come up with a plausible answer to this great mystery, we would do well to keep an open mind as to the methods used.

The above story of the mysterious Dr. Jarl and the Tibetan monks, however improbable, at least fits the bill, because it attests to a stone-raising technology that relies not on ropes, wheels and pulleys, but on purposefully created vibrations of sound. Furthermore, if Jarl's account is genuine, it seems that there may have been more to the events he witnessed than simply the use of sonics. He describes, for example, how the two hundred or so monks not directly involved with playing the instruments stood in rows eight to ten deep behind the arc of musicians, carefully following the flight path of the stone blocks as they rose up toward the cliff face. Jarl was unable to establish their true role in the proceedings, suggesting that they could have been either trainees or replacement players, or that they were engaged in the kind of "group-mind" enterprise discussed earlier in this chapter, meaning that they were effectively using some kind of psychokinesis to direct the flight of the stones.

Jarl's entire account is very sober and detailed, recording numbers, distances, angles, dimensions, and even technical specifications relating to some of the instruments themselves. As Collins says, not unreasonably in my opinion, there seems to be too much detail in this report for it to be dismissed as total fantasy.

Collins goes on to cite another account recorded by Kjellson, that of an Austrian filmmaker by the name of Linauer who also visited a Tibetan community sometime in the 1930s. Here again we have a very detailed account describing the use of custom-made instruments of sound—in this case a large gong made of gold, iron, and brass and a stringed instrument, also made of different metals and shaped something like a large mussel shell, which apparently was not played as such, but somehow worked in conjunction with the low, short-lived sound vibrations emitted by the gong. Collins suggests that the "silent" stringed instrument may have transmuted the sound of the beaten gong into the ultrasonic range, which somehow caused the effects allegedly witnessed by Linauer. He reported that when these two instruments were activated they enabled the monks to lift heavy stone blocks with just one hand and very little apparent physical exertion. Linauer was also told by the monks that similar instruments existed that could actually disintegrate physical matter. This brings to mind Gurdjieff's claim that "objective music" could do unimaginable things, freeze water, for example. More ominously, it could make flowers wither and die within minutes, cause physical aberrations to manifest rapidly in the structure of physical organisms, or even, as in the case of the warrior patriarch Joshua's assault on Jericho, cause great stone walls to disintegrate, to crash to the ground in pieces.

In chapter 6 of his book, Collins further examines evidence in sacred buildings all over the world of an extraordinary knowledge of acoustics.

In Mexico, for example, there is the nine-stepped pyramid known as the Castillo, a temple dedicated to Viracocha/Kukulcan, which is one of the main structures of the Mayan complex at Chichén Itzá in north Yucatán. If you stand at the foot of this pyramid and shout, the sound vibrations echo and transmute into an eerie shriek that emanates from the top of the building. Alternatively, if you speak in a normal voice while standing on the summit, you can be heard quite clearly by people on the ground as much as 150 meters away.

Similar strange acoustic properties have been identified in the nearby Great Ball Court, a large field 160 meters in length, flanked by two temples, where a faint whisper at one end can be heard quite easily from the opposite end.[8]

There are further examples of unusual acoustic properties in other

Mayan structures. One is the Temple at Tulum on the Yucatán coast, which gives off a long, low howling sound when the wind is at a certain velocity and blowing in a particular direction. Another intriguing example is the Temple of the Magician at Uxmal, built, according to Mayan legend, by a mysterious race of dwarfs who only had to whistle in order to make the heavy blocks of stone rise into the air.[9] If you stand at the base of this pyramid and clap your hands, the sound emerges from the top as an eerie chirping, quite unlike the original sound vibrations. At another famous site at Palenque, which consists of three principal pyramids, it is possible for three people to stand one at the top of each of them and engage in a three-way conversation.

Possibly Collins's most interesting observations concern Egypt, and in particular the King's Chamber of the Great Pyramid. Many observers have noted how voices sounded in this chamber have unusually resonant properties. It is as if this effect, among many others of course, was intentionally created. Collins suggests that this unusual sound property might have something to do with the fact that the "Pythagorean" 3–4–5 triangle is incorporated into the chamber's whole design.

This fact can be observed by describing a diagonal from one lower corner of the end wall up to the opposite top corner, which, if the baseline of the floor is included, results in a perfectly proportioned 3–4–5 triangle. The same applies to the huge block of granite incorporated in the wall immediately above the entrance to the chamber. The fact that the chamber is exactly twice as long as it is wide means that the 3–4–5 symmetry is an intrinsic feature of its whole structure.[10]

As Collins notes, this particular geometrical configuration expresses three significant harmonic proportions that together produce the keynote in a major scale, as with the notes (based on the scale of C major) D (re), E (mi), and G (so), for example, which generate the vibrations of the keynote C, i.e., the magical Do, which appears at the beginning and the end of every major scale. The combined frequencies of these three notes relate to one another in the same way as the combined ratios of the 3–4–5 triangle.

The red granite "sarcophagus" in the chamber also possesses unusual acoustic properties. When Flinders Petrie organized a team of workmen to lever one end of the sarcophagus up off the ground some twenty centimeters, so that he could take accurate measurements of its dimensions, he just happened to strike the tilted coffer with a hard implement and was impressed by the deep, resonant sound it produced, rather like a bell.

Another interesting feature of this coffer is that its external volume is exactly twice that of the internal volume. And this ratio of 2:1, as we noted, corresponds to the length of the entire chamber in relation to its width. In musical terms, of course, this proportion is highly significant, because it expresses the ratio between the two extreme notes of the major musical scale, where the last note, Do, of the octave vibrates at twice the frequency of the first note, also Do. In view of the vast number of possible variables in dimensions that the builders could have opted for, I think we can reasonably assume that these proportional symmetries did not occur simply by chance. Indeed, given the fact that the Hermetic Code was the central theme of Egyptian metaphysics, one would be extremely surprised and puzzled if such harmonic proportions were not present, in the King's Chamber or anywhere else. The whole of the Great Pyramid itself, remember, is a massive representation of the *pi* symmetry, the "trinity of octaves," so it would have been perfectly natural for the designers to have incorporated expressions of the same musical system in its most impressive internal features.

In a later chapter we shall be looking again at these lost techniques of the builders of the Giza necropolis. As I have said, I believe these methods were based on a complete understanding of the universal harmonies described by the Hermetic Code, but this in itself remains a rather abstract idea, a bit like the hippie notion of tuning in to "good vibrations" as a means of inducing a sense of well-being. There is much more to the Hermetic Code than that, however. It is a universal formula with many facets, and certain of them, as we shall see, are by no means vague or abstract, but scientific in every sense of the word. But in order to appreciate the full implications of this belief system, we first have to examine some of the wider applications of the code itself.

Finding a theory capable of unifying the whole body of our empirical knowledge into a coherent whole—a "theory of everything"—is currently the ultimate scientific goal. Now, I'm no specialist, but after a great deal of painstaking thought and deliberation, I have come to believe that the Hermetic Code could well be what we are searching for—the answer to practically all of our most fundamental questions on life and the universe. Obviously this is hardly a minor claim in the great evolutionary debate, but throughout my years of questing I have always borne in mind that I am propagating here not my ideas, but those of the enigmatic "god of wisdom" Hermes/Thoth, one of the greatest minds ever to have existed.

In order to appreciate just how far-reaching this belief system really is, we must for the time being return to the present and examine some of the fundamental discoveries of modern science. Some of the concepts about to be discussed are extraordinary to say the least, and may at first seem difficult to grasp, illogical even. But there will be no mathematics involved here—we need only have a general idea of the nature of the strange world now being described by scientists, enough to enable us to compare it with the star-strung universe of the Egyptian high priest. Therefore, as a starting point, we shall be looking into the nature of what is perhaps the most important and familiar phenomenon in existence. This is light, the "creative rays," which, Yogananda claimed, could somehow be manipulated by the trained mind of the yogi. This is saying, in effect, that there is some kind of accessible interface between mind and light.

Now these "creative rays" are actually composed of what are today known as light quanta, or photons, subatomic components classified as "virtual" particles, which means that they have no measurable mass. As we shall see in the following chapter, photons have been shown to exhibit some strange, almost ghostlike properties. And they are not alone: there are, down in the physicist's microworld, other minute components engaging in paranormal activities, in particular electrons, the particles that give all infinitesimally small atomic nuclei a hard, voluminous outer shell and hence the property of materiality as we know it. Moreover, the photon, as well as existing in the form of visible light and other rays of the electromagnetic spectrum, is also the "force carrier" of all electromagnetic interactions, which means all interactions between particles of matter, between electrons. In other words, when matter forms or decomposes by interacting with environmental conditions, it does so through the constant emission and absorption of photons. Light, therefore, as well as being a type of radiation capable of inducing in us visual sensation, is also the universal agent of change. Therefore if Yogananda's claim is correct, that these "creative rays" can somehow be influenced by the trained mind, we already have a possible explanation as to how psychokinesis might work.

Yogananda, who always followed closely the progress of modern science, noted that the word "impossible" was becoming less prominent in man's vocabulary. That was back in 1946. Since then it seems to have disappeared altogether, leaving in its stead a plethora of "improbabilities." Physicists know that, in the quantum world of subatomic particles, the impossible can and does happen. For example, it has been discovered that certain categories of virtual particles are created out of "nothing" in what most people think of as "empty space," borrowing energy from some unidentifiable cosmic storehouse only to disappear without trace nanoseconds later after paying back the energy loan. Billions of these massless entities are apparently popping into and out of existence in every cubic centimeter of space in a manner that might reasonably be described as ghostlike. So if you don't yet believe in the paranormal, either talk to a physicist or turn the page.

4

The Electron
and the Holy Ghost

Around the beginning of the twentieth century, a new era of scientific enquiry began, and with it came some startling discoveries concerning the nature of matter. Previously, classical physicists had thought of the material universe as "deterministic," that it obeyed the established Newtonian/Einsteinian laws of motion and gravity, and that all material processes could in general be predicted with experimentally verifiable accuracy. But when physicists started probing atomic structures and their components, they discovered that they behaved in random, uncontrollable ways. In order to account for the peculiar dynamics of this strange underworld, scientists developed a new kind of physics, known today as quantum mechanics.

This new science is remarkable, because its practitioners not only believe in the paranormal, but can prove experimentally that it is a reality.

It all started with the investigation of subatomic particles, the smallest entities yet detected in the universe, the components of atoms, of light and of just about everything else. Originally it was thought that they were simply particle-like points in space, but recent discoveries have shown that the "particle" observed is only the detectable trace of a much more complex entity, whose overall presence reaches far and wide.

The first hint that this was so came from investigations into the nature of light itself, which is emitted by light sources in discrete "particle packets," or quanta, of electromagnetic energy called photons.

It was noted that a thin beam of light shone through a tiny pinhole in a partition with a dark screen or photographic plate behind it creates a small circle of light on the plate. If there are two holes in the partition close together, the image on the back-plate forms two circles of light overlapping. In the area where they do overlap, however, there are intermittent dark bands, where obviously no light is present. This has been attributed to a familiar wave-mechanics phenomenon known as interference, and it shows that the light is emerging from each pinhole as waves, sometimes overlapping and reinforcing one another, and sometimes canceling one another out—hence the dark, lightless bands. Actually, the wavelike nature of light was first recognized as long ago as 1803 by the Englishman Thomas Young, using nothing more than a flame, a partition with two narrow slits and the dark backdrop upon which the pattern appeared.

Now, if single photons are fired one after another from a light gun over a given period of time, when the photographic plate is subsequently developed the interference pattern, logically, should not be there, because a single photon, presumably, cannot "interfere" with itself. Curiously, however, the interference pattern invariably does appear. The photon, it seems, can do whatever it chooses in its own surreal world. It can clap with one hand, creating interference patterns out of nothing as if in collaboration with some unseen, ghostly counterpart. Stranger still, when a photon detector is engaged to "see" what is going on when the photons emerge from the holes, the interference pattern disappears. Apparently we only have to "look" at a photon and it changes its nature completely.

Light, then, is a wavelike phenomenon. At least, that is what everyone thought until Einstein came along with a completely different interpretation of it. He formulated some equations to account for a phenomenon known as the "photoelectric effect," which is the effect of light shining on a metal surface, whereby electrons are emitted by the metal, causing an electric current to flow. His calculations proved, theoretically, that light—the photon—is a particle. This was later experimentally verified, and it was for this discovery, not the famous Theory of Relativity, that Einstein received his Nobel Prize.

Later discoveries made by physicists in the 1920s, notably those of the French aristocrat Louis de Broglie and the Austrian Erwin Schrödinger, showed that the electron, one of the fundamental components of all atoms, also has both particle and wave properties.

So what exactly is light, this ghostlike, photon entity? It is a particle with wavelike properties, a wave with particle-like properties, a mysterious, diminutive something that actually reacts when we "look" at it. If we leave it alone, it behaves like a wave, but as soon as we start to measure its movements, it flips over into particle mode. Classic abracadabra: now you "see" it, and when you do, it responds, "curls up," and changes its nature completely.

Another breakthrough experiment, again demonstrating that there is a great deal more to these wave/particles than first meets the eye, was the "twin particle" experiment conducted in 1982 by Alain Aspect and his team at the Institute of Optics in Paris.

Originally outlined by the theoretical physicist John Bell in 1964, the experiment was devised to test an apparent absurdity in the rules of quantum mechanics, first pointed out in 1935 by Einstein and two colleagues, Boris Podolsky and Nathan Rosen. Basically it concerned one of the most controversial rules of quantum theory, which says that subatomic particles are interconnected in a way that classical physicists believed was impossible.

It was discovered that certain subatomic processes result in the creation of pairs of particles with identical properties. For example, when an electron and its antimatter opposite—a positron—come into contact and annihilate one another, they coalesce into two light quanta, two photons, which then zoom off in opposite directions at the maximum speed allowed by nature—the speed of light. Quantum physics states that, irrespective of how far apart these twin quanta travel, when they are measured they will always be seen to have the very same angles of polarization. That is, at the precise moment of measurement of one or another of these particles, its twin somehow "knows" which angle is to be agreed upon. Consequently there must be some sort of instantaneous communication going on between them.

Another curious feature of quantum mechanics arises from what is known as the "uncertainty principle," which was first expressed in 1927 by the German physicist Werner Heisenberg. According to this principle, wave/particles do not have a definite position in space and time, which means that their locations can only be expressed in terms of variable statistical probabilities collated over the course of many duplicate experiments. The "uncertainty" arises from the fact that it is not possible to measure simultaneously, with a high degree of accuracy, both the position and the momentum of a moving particle. Measuring one aspect, say the position, affects the momentum, and vice versa. In other words, the very act of observation changes the primary state of the wave/particle. We noted this strange property earlier in the behavior of the photon, which, when targeted by a photon detector, switches over from wave to particle mode. The point is, in their virgin state, wave/particles do not have exact locations. Depending on how they are measured, they can manifest as a specific point, or as a fuzzy cloud of wave-like energy.

The Danish physicist Niels Bohr had a long-standing dispute with Einstein and his colleagues over the true nature of this so-called action at a distance between twin quanta. Einstein rejected the notion because it seemed to imply that there was a "superluminal" (faster than light) transference of information operating between the two coordinates, and the Theory of Special Relativity states absolutely that nothing on a material level of existence can travel faster than light. Bohr's answer to the problem, which is generally accepted by the majority of today's physicists, was that there was in fact no superluminal communication taking place, and that Einstein's error lay in viewing twin particles as being independent, self-contained phenomena. Bohr reasoned that if subatomic quanta do not really exist until a probe of some kind causes them to "curl up" and manifest one of their measurable properties, then it was meaningless to consider them as separate things. Quantum systems in their natural, "unmeasured" state are indivisible from one another, and what we observe as being apparently unrelated subatomic events are in reality in a constant and immutable state of interconnectedness, even if they are on opposite sides of the universe.

When John Bell first thought up an experiment that could verify or disprove this idea, technology hadn't yet developed to a level at which it could produce instruments with an accuracy and sensitivity sufficient to carry it through. This is why it took until the early 1980s for Alain Aspect and his group finally to take up the challenge.

The experiment involved creating a stream of twin photons by heating calcium atoms with high-energy lasers, and then allowing them to fly off in opposite directions through lengths of pipe. At the end of each pipe were special filters that deflected each twin toward either one of two polarization detectors. The accuracy of the instrumentation ensured that all the crucial stages in the experiment could be performed in so brief an instant that there would be insufficient time for even a beam of light to traverse the space between the two particles. And, sure enough, as quantum theory had predicted mathematically, each photon was always able to manifest simultaneously exactly the same angle of polarization as its twin. Consequently, physicists now believe that the connection between two such related quanta must be "nonlocal," which means that no matter how far apart they are they always remain composite parts of a single, dynamic, interconnected system.

So we know that twin photons generated from a single impact event travel out from the source of their origin at the speed of light. Now this is significant because, according to the physicist, to an observer moving at such a velocity, time and space as we perceive them would both cease to exist. As speed is increased, they say, time slows down proportionately, eventually reaching a complete standstill at the speed of light. Simultaneously, space gradually contracts, eventually into nothing, no space whatsoever. What scientists are positing here, therefore, is a dimension of existence in which space and time do not exist. This is why the photon, itself perpetually existing in this strange, "spaceless" world in which time stands still, can instantaneously "transmit" information to a twin—because the impulses carrying the data have no "space" to pass through: they are already there, so to speak.

Clearly we are talking now of an alternative reality to the one we are all familiar with, quite literally another dimension, and it is a world as curious as any found in fairy tales. In this alternative, quantum world, all entities, in moving at the speed of light, must effectively occupy, at one and the same instant, all possible locations along the line of passage. No matter how long the line as observed from a stationary frame of reference, the photon simultaneously exists everywhere along it. Like the Holy Ghost or the spirit of Muhammad or the Buddha, it is "omnipresent."

It is difficult to imagine what it would be like to see the universe through the "eyes" of the photon. There would be no distance between stars and galaxies, continents, you, or me; there would be no space, no ticking of clocks, and no aging. The moment a photon is created, say, inside a distant star, at that very same instant it could be entering the retina of an observer zillions of miles away. Its creation and annihilation is in fact one single-impact event, captured for all eternity in a frozen, timeless instant.

Similarly, if we ourselves could attain the speed of light and cross over into this other reality, theoretically we would be godlike. Just like the photon we could exist everywhere simultaneously, "visiting" remote constellations simply by focusing on them; and our conception and our death would be perceived as one and the same event, a single, permanent feature in the timeless, unfading fabric of creation. Presumably, once created, everything existing in such a dimension must exist literally forever; and what might be observed in the laboratory as, for example, the creation and annihilation of a humble photon is merely a cross-section of a much greater and more complex reality in which the observed event, which might have taken only a microsecond or two to unfold, continues to have a permanent existence independent of time.

According to Einstein, extraordinary changes would occur to a physical body if it could ever reach the light barrier. Its length, together with the length of the trajectory in space along which it were traveling, would become zero, and its mass would become infinite, expanding at right angles to the direction of motion into a vibrant sheet of wavelike energy of immeasurable size.

Now, in crossing the light barrier, any physical entity would, in effect, be transcending the fourth dimension, the line of time, and passing over to a quite different dimension existing beyond time. And if we refer to time as the fourth dimension (after the three dimensions of space: line, plane, solid), then the next in succession—what I called in my last book the plane of light—is the fifth. This is the "nonlocal" world of the photon quantum.

What is emerging here, in fact, is an overall cosmic picture of a succession of dimensions, from zero point to a line, a line to a plane, a plane to a solid, and subsequently the continuous existence of a solid along the line of time. These four different perspectives are easily recognizable, but the fifth in the ascending scale, as physicists have discovered, needs more than a little intuition to identify. In a later chapter we shall discuss in more detail these different dimensions, as they provide a convenient way of fixing our position in the cosmic scheme of things.

The point to note here is that this fantastic fifth dimension is definitely there. We know this because physicists have proven it mathematically. This is highly significant, because it raises a most interesting question: which of the two dimensions is nearer to reality, the timeless, spaceless, nonlocal world of the photon, or the world we perceive, a world of sense objects, ticking clocks, night and day, birth and death? The answer, of course, at least as far as the physicist is concerned, is that the nonlocal world of fundamental quanta is the primary reality, and that the world perceived in time by our ordinary senses is at best incomplete. This is precisely what Einstein was referring to when he wrote in a letter to the relatives of a deceased colleague, "People like us, who believe in physics, know that the distinction between past, present and future is only a stubbornly persistent illusion."[1]

What is particularly interesting about this curious nonlocal dimension of the particle physicist is that it almost perfectly matches the worldview of many so-called primitive peoples, of the aboriginal shamans, of the writers of many of the world's great scriptures and, perhaps most significantly, of the Egyptian priesthood. Remember the Hopi, whose shamans perceive only an "eternal present" and whose ceremonial dance results in those involved experiencing the collapse of the whole universe into a single event. In a similar vein we have the Egyptians of the Old Kingdom and scriptural writers from every major culture, to whom the concepts of eternity (timelessness) and infinity (spacelessness) were common fare.

As I described earlier, the psychedelic experience can result in the same kind of impression, that is, of a world in which time seems to stand still. Speaking personally, my own "extratemporal" experiences were impressive in the extreme, and it is unlikely that I will ever forget those uplifting feelings that we human beings could live forever.

Possibly such perceptions are the result of what Colin Wilson sees as right-brain, intuitive thought processes—of the kind he believes to have been used by the possessors of ancient "lunar" knowledge, which was unified and enabled people to see things as a whole. The Hopi's concept of an eternal present seems to express just such a unified worldview, in which everything in the entire universe condenses into a conceptual singularity, multiplicity becomes unity, all becomes one.

Now let's return to another strange idea that has echoes in the present, one that was first expounded at least as long ago as the time of the Greek philosophers Pythagoras and Thales. This is the notion that matter itself is "psychic," that it possesses some kind of awareness of its environment. Probably very few scholars have ever given any serious consideration to such a seemingly fanciful claim. It's a quaint idea, one might think, but we shouldn't take it to heart. And yet, curiously, some of the latest discoveries of modern science actually lend support to such a view.

We have already mentioned the peculiarly responsive behavior of the photon, which behaves like a wave when left unobserved and as a particle when targeted by a detector, and also the now proven reality of nonlocal (timeless, spaceless) interactivity between twin quanta.

In an attempt to explain the principle of nonlocality and the idea of a vast web of interconnectedness permeating the whole universe, the University of London physicist David Bohm posited the existence of what he called quantum potential. He saw this as a new kind of energy field that, like gravity, pervades the whole universe, but whose influence does not weaken with distance.

Bohm first recognized a possible indication of this quantum potential through his work on plasmas, gases comprising a high density of electrons and positive ions (atoms with a positive charge). He noticed that the electrons, once they were in plasma, began to act in concert, as if they were all part of a greater, interconnected whole. For example, if any impurities were present in the plasma, it would always realign itself and trap all foreign bodies in an exclusion zone—just as a living organism might encase poison in a boil. Bohm observed also a similar, orchestrated mass movement of electrons in metals and superconductors, with each one acting as if it "knew" what countless billions of others were about to do. According to Bohm, particles act in this way through the influence of the quantum potential, a subquantum force matrix that somehow coordinates the movement of the whole.

It appears that when plasmas are rejecting impure substances and regenerating themselves, they look very similar to swirling masses of well-organized protoplasm. This curious "organic" quality led Bohm to comment that he often had the impression that the electron sea was, in a sense, "alive." He possibly did not intend this to be taken too literally, that the electron mass was living in the same way as an amoeba, but the evident highly coordinated symmetries of the plasma convinced him that the electrons were responding to one of many "intelligent" orders implicit in the fabric of the universe. He believed that order exists in many different degrees, some forms being much more ordered than others, and that as a consequence the things we see as disordered at our ordinary levels of perception may in fact be perfectly ordered when viewed in a more objective way.

To illustrate this point, imagine yourself as a microcosmic visitor in a living cell, observing amid a writhing sea of biomolecules—proteins, enzymes, amino acids, and the like—all busy exchanging energies, whizzing past you in a flurry of hyperactivity. What you would see might appear to be virtual chaos, a seething marketplace full of eager bargain-hunters, pushing, gathering in random groups, shouting, haggling. But, in fact, all this frenzied activity, appearing on the face of it to be an unending display of random physical actions, is totally governed by the hidden DNA of the cell, possibly one of the most organized and beautifully proportioned structures in the entire universe, and producing, as a direct result of the cell's activity, a greater organism of an infinitely higher order.

So these electron symmetries, which Bohm called plasmons, appear to be following hidden instructions encoded somewhere in the subquantum fabric of the universe. But even where we observe no apparent orchestrated activity, where masses of electrons seem to be acting randomly, we may simply be trying to view them on the wrong scale—rather like our microcosmic onlooker in the biochemical marketplace of the cell.

Bohm was ultimately to conclude that the ordinary world as seen through orthodox scientific experimentation is really an illusion, something like a holographic image, and that somewhere behind this lies a much deeper and more meaningful level of reality—the holographic "film," as it were, from which the image originates. This metaphor of the universe as a living hologram subsequently became the central theme of Bohm's investigations, which have been summarized by Michael Talbot in his book *The Holographic Universe*. We can take another look at the wider implications of this important concept later.

Bohm's views on consciousness in relation to matter are also interesting. He believed that consciousness itself is actually a subtle, highly rarefied form of matter and that forms of intelligence exist, in correspondingly different degrees, in all kinds of material substances. "The ability of form to be active," he said, "is the most characteristic feature of mind, and we have something that is mindlike already with the electron."[2]

As it happens, Gurdjieff and Ouspensky were saying much the same thing in the early part of the twentieth century, that everything, including all our finer thoughts and aspirations, has a material existence and could, theoretically, be weighed and measured. On the subject of matter as we know it, Gurdjieff had this to say: "In addition to its cosmic properties, every substance also possesses psychic properties, that is, a certain degree of intelligence."[3]

Do these observations seem at all familiar? They sound decidedly "Greek" to me. Bohm's electrons, negatively charged wave/particles that orbit the nuclei of atoms at velocities approaching the speed of light, are what give matter its substance, its apparent solidity. And if electrons exhibit "the most characteristic feature of mind," then this means that the Greeks were right all along and that all material things are endowed with "psychic" properties.

In fact, Bohm then took this highly mystical worldview a giant leap farther by suggesting that not only are "inanimate" objects like rocks and stones in some way alive and intelligent, but so too is all energy, all time, all space—everything. As we noted earlier, Sri Aurobindo expressed a similar view when he said that if there were a single point in the universe that were not conscious, the whole universe itself would be unconscious.

The principle of nonlocal interconnectedness is hereby taken to the absolute limit, where even so-called empty space is seen to be full of meaning, brimming with an infinite store of primordial intelligence, the underlying formative matrix for everything existing, including ourselves. Thus all the phenomena we observe in the physical universe are simply "ripples" on the surface of an unimaginably vast ocean of deeper meaning. This hidden world Bohm called the implicate or enfolded order, the subsurface dimension that gives rise to the phenomena we observe with our senses, in the explicate, unfolded order. So the manifestations of all forms are the product of endless enfoldings and unfoldings between these two very different but mutually interconnected dimensions.

In this way a wave/particle, like an electron, is described not as one thing, but as a nebulous stream of interchangeable energies enfolded throughout the whole of space. When it is measured by an investigator, what is observed is merely one property of the "greater electron," which has simply responded to some probe or other by unfolding into the explicate order.

Obviously the ancient Greeks would have known nothing of the strange properties of the subatomic particle. Nevertheless, they still somehow managed to establish a view on the mindlike nature of materiality that accords with the latest discoveries of modern science. How? Was it a lucky guess? Did someone perhaps tell them? Or was it just plain old-fashioned intuition? Of the three possibilities, I suspect that the first is the least likely. For reasons that will become clear a little later, I am inclined to believe that the Greeks received this wisdom from their predecessors, but that intuition played a large part in their understanding of the teachings they inherited.

So what else is "Greek" in this present era of scientific discovery, with its particle consciousness, photon "telepathy," and so forth? Is there any other knowledge that these ancient peoples possessed that might be relevant to this enquiry? Indeed there is: there is the knowledge that they received directly from the Egyptians in the form of the Hermetic Code, which says that everything in this universe manifests strictly according to musical principles.

Once again we can see how strangely "modern" is this view, because scientists themselves are now speaking more and more in terms of a musical universe that endlessly vibrates, and of physical phenomena all possessing unique resonances of their own.

For example, in his book *Other Worlds,* Paul Davies describes the way electrons orbit the nuclei of atoms in a regular order, whereby only stationary patterns will occur. He compares the phenomenon to the standing wave-pattern of air in a particular set of organ pipes, where only certain established notes are permitted because the patterns of air-waves must fit into the geometry of the pipes. Similarly, only certain "notes" (frequencies, energies) are accommodated by the atom. When transitions occur between the normal energy levels, electrons emit characteristic colors—streams of photons—and these are the visual evidence of what Davies calls "this subatomic music." He continues:

> We can therefore regard the spectrum of light from an atom as similar to the pattern of sound of a musical instrument. Each instrument produces a characteristic sound, and just as the timbre of a violin differs markedly from that of a drum or a clarinet, so the color mixture of light from a hydrogen atom is characteristically distinct from the spectrum of a carbon or uranium atom. In both cases there is a deep association between the internal vibrations (oscillating membranes, undulating electron waves) and the external waves (sound, light).[4]

There are other musical relationships between atoms and their components. For example, all atoms are members of a whole, integrated family, ranging from the lightest, hydrogen, with one electron tracing a lone orbit around its nucleus, to the densest, heavily radioactive atoms, which have many electrons orbiting the nucleus in seven permitted energy levels. Remember that there are seven successive "energy levels" in the major musical scale. Obviously the eighth, transcendental "note" of this fundamental atomic scale is the whole phenomenon, consisting of all atoms everywhere.

Further, a recently developed classification system known as the theory of quantum chromodynamics suggests that beneath the materiality of the atom there are other essentially musical symphonies being played by nature. Scientists are currently classifying a certain category of subatomic particles according to a system known as the eightfold way. The theory is so called because it puts certain routinely observed "particle molecules" known as baryons, pions, and mesons together in families of eight. The term was originally coined by the American physicist Murray Gell-Mann and was intended as a pun. He was apparently familiar with the "eightfold path to enlightenment" devised by the

Buddha, and presumably felt that the name would add a lighter note to his complex mathematical theory. Doubtless the idea that the Buddha's belief system is in any way scientific would make Gell-Mann's toes curl. But, being unashamedly what the science writer Richard Morris has referred to as "one of those deluded mystics who manage to see parallels between theories in physics and ideas associated with Eastern mysticism," I would suggest that this is precisely the case, that it is no mere coincidence that the Buddha's musical interpretation of reality should so easily and naturally blend in with the foremost ideas of today's scientists. The "eightfold way" of the Buddha is a variation on the Hermetic Code, and like the "eight steps of learning" of his Chinese counterpart Confucius, it was founded on the idea that the whole universe is an essentially musical structure and that to realize this, to tune in to this fundamental reality, one had to conform to the laws and forces controlling it.

Thus, if we look closer at this chromodynamic system of classification, we shall see a quite familiar pattern emerge.

There are supposedly eight low-mass baryon wave/particles making one octet, eight pions forming a second octet and eight vector-mesons making a third—twenty-four in all. Now, this same family of particles also comprises, in addition to the octets, a complex triplet. This means that each of the eight particles in an octet is also a triplet, made up of three smaller particles, which Gell-Mann called "quarks." As we see, the structure of each octet (or octave) of triplets is identical to the symmetry of the I Ching, with its eight trigrams. And there is more. Gell-Mann's theory originally called for three kinds of quarks, called up, down, and strange—a subatomic "trigram." But, then, to these were subsequently added three more types of quark, called charm, bottom and top. Enter the hexagram. All we need now to complete the picture is the number 64. It would be highly fitting if we could find it, because sixty-four is not only the number of hexagrams in the I Ching, it is also the number comprising the council of Brahmins who, according to legend, foretold of the impending birth of the Buddha. In fact, as I pointed out in *The Infinite Harmony,* this particular number has surfaced not in quantum chromodynamics, but in what is known in physics as superstring theory.

The central idea in superstring theory is that subatomic wave/particles are in reality infinitesimally small strings made of space. These strings vibrate endlessly over an infinite range of frequencies, and their interactions give rise to the observed characteristics of all known particles. You really can't get more intellectually obscure than the theory of superstrings, and I am personally completely baffled by it, involving as it does no less than ten different dimensions (three of space, one of time, and six of God-knows-what) and a system of higher mathematics guaranteed to make the layman's eyes glaze over in seconds. But no matter; all we need to know here is that this incredibly complicated system has created a superstring, out of nothing but space, that has precisely 64 degrees of movement associated with it. This supersymmetric system can apparently account for all subatomic quanta, and is capable, says the science writer Timothy Ferris, of "drawing all matter into an elegant picture in which particles' attributes are seen as the vibrations of strings, like notes struck on Pythagoras' lyre."[5]

So we're back to Pythagoras again, the original philosopher, a contemporary of the Buddha born five and a half centuries before Christ, who taught that everything in the universe obeys musical laws and who, like the yogis of India, believed that matter was "psychic." And both of these ideas, as we have seen, have now gained a metaphysical foothold in the mind of the modern scientist.

Now, if consciousness is material in some way—as Gurdjieff and Bohm both believed—and if matter is conscious, though on an entirely different scale, then could the higher possibly influence the lower, and vice versa? Pythagoras would very likely say yes to the former proposition, possibly citing the mysterious powers of Orpheus and Amphion as examples. But he would also, being what today might be called a natural mystic, have believed that the psychic presence in matter could indeed influence human beings. People today of a sensitive or intuitive inclination often feel that nature speaks to us in many different ways.

Mountains and forests, for example, as many people instinctively know, have a particularly powerful presence. So too do many ancient artifacts, such as the Great Pyramid or the Taj Mahal, the Cathedral of Notre Dame, or a statue like the Sphinx.

Ouspensky recognized a similar close relationship between himself and nature. He describes one of his drug-induced experiences in his second major classic, *A New Model of the Universe:* "Everything was living, everything was conscious of itself. Everything spoke to me and could speak to everything. Particularly interesting were the houses and other buildings that I passed, especially the old houses. They were living things, full of thoughts, feelings, moods and memories. The people who lived in them were their thoughts, feelings, moods."[6]

In another passage, he preempts the modern physicist by describing the world he was seeing as "a world of very complicated mathematical relations": "this means a world in which everything is connected, in which nothing exists separately and in which at the same time the relations between things have a real existence apart from the things themselves; or possibly, 'things' do not exist and only relations exist."[7]

Sri Aurobindo saw the world in exactly the same way. In his view, all apparent separateness on the physical plane is simply an illusion. In the state of enlightenment, he said, the unity of everything is perceived as a living reality, but as one descends from the higher to the lower states of consciousness, a progressive "law of fragmentation" takes over and "things" appear once more as isolated, separate entities.

And science, of course, now supports this view. As we have seen, all subatomic particles are also waves of different frequencies, and this means that everything is composed of a vast, interconnected web of interference patterns. Talbot, in *The Holographic Universe,* suggests that our brains mathematically construct this so-called objective reality by decoding these varying frequencies that are really projections from another dimension existing beyond space and time. So perhaps the great ocean of waves and frequencies "out there" looks solid and real to us only because our brains automatically reprogram all this "fluid" information into the familiar form of the sense objects making up our world. In reality, however, everything is a vast sea of highly resonant interference patterns. The sun and stars and the planet we live on, the Great Pyramid and the Sphinx, even the brain itself—all these physical structures are in essence composed of overlapping waves.

In the last chapter we discussed the work of Robert Jahn and Brenda Dunne, whose experiments with the random-event generator and the "pinball" machine provided compelling evidence for psychokinesis. Having found evidence of this ability in a large proportion of their subjects, they came to some interesting conclusions concerning the possible nature of such a process. They proposed that since all physical phenomena possess a particle/wave duality, then perhaps consciousness does too. When in a particle-like state, consciousness would be localized inside the skull, but when in a wave mode, like all waves, it can produce effects at a distance.

In a similar vein, though not in relation to psychokinesis, the Cambridge mathematician Roger Penrose has also considered the effects of quantum processes in respect of the workings of the human mind. When speaking of "action at a distance" between twin particles (non-local quantum correlations), he suggests that such phenomena could be involved in conscious thought processes over large regions of the brain itself, and that perhaps there is a direct relation between a "highly coherent quantum state" and a correspondingly high degree of awareness.

Jahn and Dunne have suggested that phenomena themselves are actually products of the combined interference patterns created by the wave motions of matter and the wavelike aspect of consciousness. They believe that psychokinesis occurs through an exchange of certain information between physical things and the human mind, not as a single directional flow from one to the other, but rather as a mutually interacting "resonance" operating between the two. These resonances sound something like the relations between "things" described by Ouspensky in the passage quoted earlier. Significantly Jahn and Dunne reported that the more successful volunteers often described a sensation of feeling "in tune" with the device.

Again, this is precisely what Indian philosophers and yogis have been saying since the dawn of their culture, that matter is responsive and that it is composed of resonating interference patterns, principally those of light itself. In his major work *On Yoga*, Sri Aurobindo describes a sphere of existence beyond space and time comprising a "multicolored infinity of vibrations," of waves. Physical reality, he said, is simply a "mass of stable light"[8]—which is precisely the conclusion I came to way back when I was experimenting with various hallucinogens. But all of this "stable light," according to the yogi, also possesses a measured degree of consciousness. This is apparently how yogi masters are able to influence the physical world: they have perfected a way of making direct contact with its rudimentary consciousness. Yogananda says much the same thing in his book *Autobiography of a Yogi*—that matter is simply "an undifferentiated mass of light." The "law of miracles," he said, "is operable by any man who has realized that the essence of creation is light."[9] So light has a very special place in the belief system of Hindus, which of course is why their most important annual festival—Diwali—is known as the festival of light. In fact, Hindus, Buddhists, and Eastern philosophers in general all emphasize the importance of light in their cosmological view of the world. Tune in to it, they say, and a whole new world unfolds. And so it would, for science tells us that light, the photon quantum, exists and operates in a timeless, spaceless, nonlocal realm. This, in my view, is the "eternal" world of the Hopi shaman, who can hold a "spaceless" universe virtually in the palm of his hand; the "infinite" world of the Egyptian priesthood, who taught that the soul of the god-king can exist for "eternity"; the "heaven" identified by all the great revelationists in history, by people who have succeeded in glimpsing beyond the veil and bequeathed to us their illuminating testimonies of the extraordinary things they witnessed.

And, clearly, the prime mover in this nonlocal dimension is light, the Holy Ghost. In this chapter we have seen how the modern scientist interprets this important phenomenon. In the following section we shall see what the primitive dreamers of former ages had to say about it.

5

Further Light

The phenomenon of light is celebrated in all of the major religions. If you read your Bible, Koran, or Upanishads, you will see that it is always spoken of in glowing terms.

In *Autobiography of a Yogi,* Yogananda quotes freely from the Hebrew and Christian Scriptures to emphasize the importance of light with respect to mankind's innate spiritual quest. He notes, for example, that God's very first command was, "Let there be light" (Genesis 1:3). He also quotes from Matthew 6:22, a verse that runs close to Jahn and Dunne's idea of consciousness operating in tune with the wavelike nature of reality: "The light of the body is in the eye. If, therefore, thine eye be single, thy whole body shall be full of light."

In the Koran, in the chapter entitled "Light," Muhammad uses the term to describe the creative power of Allah: "Allah is the light of the heavens and the earth. . . . Light upon light, Allah guideth unto His Light whom He will."

Similarly, in ancient Persia the principal god, Ahura Mazda, was associated specifically with light. It is said that when the prophet Zoroaster achieved enlightenment, it was through the agency of a spirit that led him to the formless light of the creator. The alchemical element fire was so sacred to the Zoroastrians because at night it was a continuous source of the creative light of Ahura Mazda.

In early Greece, the sun and its light were revered in the form of the gods Helios and Apollo. According to myth, Helios, father of the hero Phaeton, had the ability to "see all things" and was enthroned amid rainbows (light) and the hours, attended by the four seasons. An almost identical description of this all-seeing creator is to be found in Revelation, where St. John depicts God sitting on his throne in heaven, encircled by the colors of the rainbow ("seven lamps of fire," Revelation 4:5).

The traditional idea of a sun-king is an important one, appearing in cultures all around the world. In ancient Egypt, a pharaoh's name often ended in the suffix *Re* or *Ra* ("sun") to indicate his divine status, as in Menkaura (Greek Mycerinus), alleged builder of the third Pyramid of Giza.

Similarly, in Central and South America, Aztec, Inca, and Maya legends all speak of an ancient god, a cosmic creator who appeared from the eastern sea soon after a major catastrophe had obscured the sun. Known by various names—Kontiki, Viracocha, Kukulcan, Quetzalcoatl—this god is said to have brought back the sun and its light, and with it civilization and a new way of being.

This ancient theme of light after darkness is the key to virtually all midwinter festivals in the northern hemisphere. Like the Zoroastrians, the peoples of Bronze Age Europe used fire burning through the night to invoke the return of the sun, its warmth, and its light.

In numerous other long-standing traditions, teachers, priests, and shamans have consistently attributed to the sun and its light, or the stars and their light, divine or supernatural significance. The Egyptian priest-astronomers, however, were the first to place the sun (Ra) at the center of a cosmological belief system. Now this, I would suggest, was not simply an abstract notion of giving thanks and praise to the giver of life. Nor was it just the sun itself that was of prime importance, but rather its light. The Great Pyramid was known to the early

Egyptians as Khuti, "The Lights," not only because of the dazzling
reflective properties of its original, highly polished casing of white tura
limestone, but because light itself was the key to their entire system
of belief. This is an important and until now unrecognized feature of
Egyptian metaphysics, and it represents something of a departure from
recent suggestions that the Egyptian religion was either a "star cult"
or a "sun cult." In reality, it was neither and it was both, the common
feature being light itself, which is emitted by all stars. In later dynasties, major temples were carefully constructed along axes aligned with
the first rays of the rising sun on specific solstices or equinoxes, a particularly striking example of which is the Temple of the Sun—Ammon
Ra—at Karnak. Ammon, or Amun, the "Hidden One," was said to be
the power behind the sun (that is, its light) that keeps the balance of
life and creation in the universe.

Given the fact that the sun is the dominant star in our sky, it seems
perfectly natural that early man should have revered it in one form or
another. But the Egyptian worldview, that mankind's future "spiritual" evolution is in some way connected with the starry world, the sun
included, was not simply idol worship based on blind faith or primitive
superstition. On the contrary, it was a carefully thought out scientific
theory, the theory of transcendental evolution, that holds that life, or
consciousness, has the potential to evolve, through the systematic application of the principles of the Hermetic Code, into higher states of
being, into cosmic scales of awareness.

The ancient Egyptians, I believe, saw consciousness, or "spirituality," like everything else, as a form of resonance operating over a whole
range of hermetically related frequencies. So the more harmonious the
mind becomes, the finer and more penetrating are the frequencies at
which it operates and therefore the higher the scale of its psychological or spiritual existence. And, to the Egyptians of the early dynasties,
this "higher scale," as we have previously noted in the astronomical
alignments of the Sphinx and the Pyramids of Giza, seems to have
been closely associated with the stars, with Orion's Belt, with Sirius,

the constellation of Leo, and in fact every other major constellation in the great wheel of precession, whose immensely long cycle, as we noted in a previous chapter, is encoded in some of mankind's most ancient myths.

In this way we can see that the Giza necropolis is not simply an old, worn-out signpost showing the way to Tombstone. It is, in effect, a giant cosmic pointer, one that naturally directs the attention of all contemplating its mysteries skyward, toward the higher, stellar scale of existence. More than that, incorporated within the dimensions of the Great Pyramid is the sophisticated mathematical relationship known as *pi,* which is first and foremost an expression of the Hermetic Code, the code by which all evolution proceeds, from DNA upward, to the conscious mind of mankind—and beyond. And then, significantly, we have the old Egyptian name for the Great Pyramid: Khuti, "The Lights." In my view we are being told here, in clear and precise terms, that the vehicle by which consciousness can transcend onto this higher scale is none other than light itself. Light and consciousness, in other words, are complementary aspects of the ongoing evolutionary process of creating higher and more sophisticated forms of "life."

We can describe such a process very easily in musical or hermetic terms. We know that there are seven fundamental notes in an octave—Do, re, mi, fa, so, la, ti—and that there are seven fundamental color frequencies in the spectrum of light: red, orange, yellow, green, blue, indigo, and violet. The eighth note, Do, of an octave is a repeat, at a higher pitch or frequency, of the first base note, also Do. Being in turn the first base note of the next evolving octave, this eighth note automatically becomes the medium through which the impetus created by the given series of notes transcends onto the scale above. In the same way, the eighth "note" in the visible scale of the electromagnetic spectrum—white light, Yogananda's "creative rays"—can also be envisaged as having transcendental properties, being the medium through which evolutionary consciousness can move on to a higher scale and so realize its optimum potential. Thus we might say that this quantum leap, from the line of time, the fourth dimension, to the plane of light, the fifth dimension, represents one fundamental octave of evolution, the magical transition from the lower Darwinian scale of existence up into the infinitely higher scale, the home of the mythical gods of the Egyptian pantheon.

The rather startling implication is that the ancients' vision of an eternal sphere—"heaven," or the starry world—and the nonlocal plane of light described by today's physicists, are each referring to one and the same level of reality. In other words, the mysterious quantum universe in which light exists, with its infinite web of instantaneous information highways and its zero dimensionality, has been "seen" at first hand and subsequently described by the priesthood of a civilization that existed five thousand years ago.

Obviously this is a personal and somewhat speculative interpretation of the religion of the Egyptians. I'm out on a limb, so to speak. But years of reflecting on the overall effect of Egyptian metaphysics on the human race, on the way it has permeated through to every major religion in history and even, as we have seen, into the disciplined mind of the modern scientist, has convinced me that the orthodox view of this ancient culture is unjustifiably restricted and ungenerous. These people, I believe, knew as much about life, the human psyche, and the universe at large as we do now. I would go even further and suggest that they may have understood a great deal more, albeit in a different way. Clearly there were men of unparalleled genius living back then in Egypt. We have unquestionable proof of this: we have the Great Pyramid, "The Lights," the greatest and most complex building ever constructed in stone, we have the detailed precessional data encoded in their myths and, perhaps most importantly, we have the Hermetic Code, the "theory of everything."

Having established that the Egyptian religion was, in fact, a theory of evolution, we can go beyond theory and look for a practical application of the principles involved. To do this, we need to focus on what is known as the "pyramid ritual." According to Bauval and Gilbert, this was the initiation ceremony conducted inside the Great Pyramid that was designed to assist the soul of the dead pharaoh on its transcendental journey to the stars.

Basically, they believe that it involved two ceremonies. The first of these took place in the Queen's Chamber, whose southern shaft is now believed to have been targeted on Sirius—the star of the goddess Isis—as it culminated at the meridian in 2550 BCE. Here the son of the dead pharaoh traditionally performed a ritual called "opening the mouth." This was carried out with an implement made of meteoric iron—the sacred adze—that was used to pierce the embalmed mouth of the mummy, an act that was supposed to restore new life to the pharaoh. After this, the second ceremony was performed in the King's Chamber, whereby the soul of the pharaoh, now charged with a new kind of life force, was freed to fly up the southern shaft, which Bauval suggests was originally targeted on Zeta Orionis, the star of Osiris.

On first impression one might think that these two rituals, however broad and imaginative in concept, served no practical purpose whatsoever, being merely an embellishment of an abstract notion of a life after death. But in my view this whole ceremony was merely an exemplar and as such was not designed exclusively for the liberation of the soul of one dead pharaoh. The ritual was for all initiates and could be performed by anyone virtually anywhere, with or without a Great Pyramid. It is a detailed description of a simple but highly effective alchemical "trick" performed by the creative mind, whereby consciousness is put first into a fundamentally passive mode, symbolized by the feminine aspect of universal creation, the goddess Isis. This does not mean a mind that is idly passive, like one absorbed in, say, watching television, but one that is consciously and actively receptive—a passive force in tune with the greater cosmos, as opposed to an empty receptacle soaking up images from a screen. A mind properly controlled in this way automatically becomes a fertile place in which new perceptions, new concepts, can germinate and come to fruition. The entirely new, active mode of thought engendered in this process is symbolized in Egyptian ritual by the transcendental journey made by the soul to the home of the god Osiris. This simply represents a new level of consciousness, a higher degree of cosmic awareness. Intuition, one suspects, is one of its manifestations.

So the pyramid ritual is a symbolic description of the process of transcendental evolution, the process by which the human mind ascends to a higher scale of existence in an essentially musically structured universe. This, I believe, is the key message of the Egyptian mysteries, and it is precisely this same idea that lies at the root of the world's major religions, all of which were set in motion by men who fully understood, and lived by, the principle of psychological harmony. This is why they repeatedly emphasized the importance of composing the mind in a certain way, through meditation, contemplation, prayer that ends in a silent gesture of submission, or whatever. These exercises were designed to create a state in which the mind is open—like the pharaoh's mouth—and so receptive to greater cosmic forces. And what are these forces? Well, according to such as the Egyptians, the Greeks, and the yogis of India, these forces are, in fact, electromagnetic.

The vehicle of all our visual impressions is, of course, light itself. A passive mind with a visual cortex focused on its environment absorbs light quanta by the billion. This is normal; it is what our retinas are designed to do. But possibly a mind operating not at an ordinary level of awareness but in a state of "optimum psychological resonance" might be capable not only of absorbing external stimuli—light quanta, impressions, and so on—but also of assimilating them in a vastly more productive and effective way than is normally possible. This, I would suggest, is the true basis of medieval alchemy, a kind of "biometaphysical" assimilation of impressions, whereby a balanced mixture of psychological elements is fused together to make spiritual "gold." Put simply, this is the process of the "transmutation" of one's impressions into finer, much more precious "substances," namely concepts. We should note here that the Arabic word *alchemy* has its roots in Egypt, which was known in old Arabia as the Land of the Chems—the Egyptians.

Alchemy means, literally, "the Egyptian way." It is also the origin of our own word *chemistry,* a philological legacy that, if nothing else, demonstrates just how potent and far-reaching these Egyptian influences can be.

It may seem something of an oversimplification of this elaborate pyramid ritual to say that it is nothing more than a description of a single alchemical process, a simple trick of the mind. But then it has frequently been acknowledged, particularly among scientists, that the most elegant ideas and theories are often the most simple, sometimes so much so that, once known and understood, they become obvious. Perhaps, then, these Egyptian mysteries, whose purpose must have been obvious to the people who created them, can also be understood without recourse to masses of technical data—simply by using basic common sense.

In fact, the "trick" in question is straightforward only in theory. In practice, it can be a most difficult thing to accomplish, at least for sustained periods. Yogis say that it takes years to master the art completely, to learn to compose the mind for periods long enough for new concepts to take root within it. But this is leading on to wider psychological issues involving detailed systems of self-discipline, the development of powers of concentration, of the will, and so on. We can return to this question in a later chapter. For the present, it is the theory itself that is of primary concern and in particular the idea that "creative light rays"—the Egyptian *khuti*—play a fundamental part in the whole musically structured process of evolution.

Khuti itself—the Great Pyramid—was known in Chaldea as *Urim middin,* which means "Lights-measures." The name is significant because it suggests that the monument was something more than an elite place of initiation or a mere symbol for a solar cult: that in fact the Egyptians had quite literally encoded within it measurable data relating to the phenomenon of light.

When one thinks of "lights-measures," the first thing that comes to mind is the speed at which light travels, which is about 300,000 kilometers per second. Physicists regard this speed as an absolute physical law; nothing, they say, can travel faster than light. In Einstein and Herman Minkowski's famous equation relating energy to matter, E = mc^2, c^2 is the constant velocity of light multiplied by itself. The square of the constant, therefore, is an important number in theoretical physics because, when it is multiplied by a factor of m—the mass of a given thing—it gives a value for the amount of nuclear energy latent within it.

So, to the obvious question: is it possible that this particular "lights-measure"—the velocity of light—is encoded somewhere within the design of the Great Pyramid, or in texts relating to it, or in the Giza necropolis as a whole? Most academics would no doubt regard such a notion with the same kind of derision that theories about Atlantis and holy space-invaders have engendered over the years. After all, these people supposedly hadn't even invented the wheel back in 2500 BCE, so how could they possibly have had any inkling of scientific absolutes?

Despite such an obvious contradiction, we can be reasonably sure that the Egyptians, like today's physicists, regarded light as the ultimate phenomenon, a yardstick by which all things could be measured. This is not to say that they ever attempted to calculate the velocity of moving objects in terms of distance and time, but that they measured or perceived things—light included—as forms of resonance obeying musical laws. These are laws, remember, that were expressed symbolically, not only through the *pi* convention but also through myth, in particular the myth of the original pantheon of eight gods who, it is said, all appeared simultaneously out of an "island of flame," an island of light.

Light, therefore, seems to have been viewed as the most vibrant of all phenomena, an octave of resonance operating at absolute or optimum frequencies—in effect a musical constant.

In the light octave, as in any other, there are eight fundamental "notes"—or colors—the seven primary ones and white. The combined harmonic value of these eight "notes" corresponds to the overall frequency at which light resonates—the constant rate. This concept of an "eight-note constant" is particularly interesting, because if we follow the example of Minkowski and multiply it by itself we end up with a value for the square of the constant of sixty-four "notes." This is significant because the Greeks associated the Great Pyramid with another interesting number relating to an "Egyptian" system known as the Magic Square of Mercury (Mercury is a Romanized name for Hermes/Thoth). This is the number 2,080, the sum of all the factors from 1 to 64. The "Minkowski shuffle," it seems, is a very old trick indeed.

I'm no scientist, and higher mathematics gives me vertigo, but it seems to me, as I have stated previously, that the modern quantum view of a nonlocal universe in which light, the omnipresent Holy Ghost, is the prime mover, was at the very least intuitively perceived by the metaphysicians of ancient Egypt. Let's say they somehow attained a higher level of consciousness, which enabled them to tune in to the quantum field, to penetrate the plane of light, where everything, as it were, resonates at the constant rate. (Actually there may be certain evolutionary processes operating in the universe that resonate at the constant rate squared, and even at frequencies infinitely higher—but that's another chapter in an ongoing saga. We can come back to this idea at a later stage.)

Of course, there is one fundamental difference between the old Egyptian science of "lights-measures" and the modern quantum description of light: the former science not only encompasses "values" for the constant, and the square of the constant, it also recognizes the phenomenon of light as an essentially musical or hermetic manifestation, an octave. More than that, this visible spectrum of seven combined frequencies also has three principal aspects connected with it, which we identify as the three "primary" colors. So light is, in effect, a "triple-octave" structure; it is an electromagnetic manifestation of the *pi* symmetry, the Hermetic Code.

It is now generally accepted that this same code is the basic blueprint of the geometry of the Great Pyramid, "The Lights," whose height (481.3949 feet) stands in relation to its base perimeter (3023.16 feet) as the radius of a circle stands in relation to its circumference. Therefore, if we multiply the height of the Great Pyramid by 2*pi*, we obtain a precise value for the monument's base perimeter: 481.3949 × 3.14 × 2 = 3023.16 feet.

It so happens that the value of *pi* is incorporated in the dimensions of another unique monument, also a pyramid, located on the opposite side of the Atlantic. This is the Pyramid of the Sun at Teotihuacán in Mexico, which is also, in my view, dedicated to the phenomenon of light. Whereas the angle of slope of the Great Pyramid is 51 degrees, 51 minutes, the angle of slope of the Pyramid of the Sun is approximately 43.5 degrees. The base perimeter is 2932.8 feet and its height is (or was) approximately 233.5 feet. Obviously the 2*pi* relationship cannot be applied here, because the Pyramid of the Sun has a much gentler angle of slope than that of the Great Pyramid. But if we substitute 4*pi* into the equation and multiply it by the height of the Pyramid of the Sun, we once again obtain an accurate value for the measurement of its perimeter: 233.5 feet × 3.14 × 2 = 2932.76 feet.[1]

What we have here, then, are two quite distinct "solar" cultures, separated by a great expanse of ocean and (possibly) time: the builders of both went to a great deal of trouble to construct gigantic pyramids with dimensions and proportions indicating a knowledge of the *pi* relationship.

And there is more. This same ratio has also been identified very recently in the structural dimensions of other important megalithic constructions of the ancient world, namely Stonehenge and another example of a pyramid, Silbury Hill in Wiltshire, both in southern England.

In his book *Thoth, Architect of the Universe*, Ralph Ellis points out that the two central pillars in the "inner horseshoe" formation at Stonehenge, the two "trilithons," which were originally capped by a slightly overhanging, curved lintel, would in their original state have given a graphic representation of the Greek letter *pi*. There is nothing particularly remarkable about this in itself. As Ellis himself acknowledges, there are "Arcs de Triomphes" in many ancient and modern cities. But, according to Ellis, the dimensions of the two trilithons, the most finely dressed stones of the entire monument, confirm mathematically that the *pi* symmetry was recognized by its designer.

Ellis uses as his units of measure what he calls the Zil yard and the Zil foot. A Zil yard, equivalent in length to the Old Saxon yard, is 1.004 meters. In the 1960s, Alexander Thom, a professor of engineering at Oxford University, established that Neolithic sites such as Stonehenge, Avebury, and many others in Western Europe had been designed using a unit he called the Megalithic yard, equivalent to 0.83 meters. But Ellis believes that his ancient measures are equally valid factors in the geometry and dimensions of such monuments as Stonehenge. His line of reasoning in this respect is a little too involved to detail here, but if we take the Zil foot as being one third of an Old Saxon yard and use it to measure the height of the two trilithons and the distance between the centers of them, we find that all-too-familiar ratio: height, 22 Zil feet, width, 7.

Ellis shows further that this same ratio is incorporated in the dimensions of the step pyramid of Silbury Hill, which is now believed to have been constructed in the same era as the Pharaoh Zoser's Third Dynasty step pyramid at Saqqarah. Whereas the Great Pyramid has a perimeter equal to $2 \times pi \times$ height, and the perimeter of the Pyramid of the Sun in Mexico is equal to $4 \times pi \times$ height, the Silbury pyramid's perimeter, whose angle of slope is exactly 30 degrees, is equal to $3.5 \times pi \times$ height.[2]

We thus have four extremely ancient structures from widely separated cultures, three of them giant pyramids, the dimensions and proportions of which all indicate a knowledge on the part of their designers of the *pi* relationship. In the case of Stonehenge and Silbury Hill, apart from their dimensions or astronomical associations, they stand mute, unsupported by any long-standing myths relating to their creators. Fortunately this is not the case with the cultures of Egypt and ancient America, whose legends abound with stories describing the godlike qualities of the creators of their extant architectural masterpieces. And when we begin to compare the myths from the Americas and from Egypt we find that the pyramid structure is by no means the only common factor. Graham Hancock has already pointed out the similarity in the facial features of the god Viracocha depicted in sculptures in South America, and Osiris in Egypt, both of whom are portrayed as bearded, light-skinned Caucasians. In addition, we have the evidence of the structures themselves, all of which have been constructed out of blocks of immense proportions. But there is also another significant common factor in the traditions of these two cultures, and this is light.

We have heard how Osiris and his resilient band of survivors materialized simultaneously from an "island of flame." And the most characteristic feature of a flame, of course, is the light it generates, which again reminds us of the old Egyptian name for the Great Pyramid: *khuti*—"The Lights."

In some of the legends of the Maya, the god Viracocha is said to have landed with a number of companions on the shores of the eastern sea following some kind of global catastrophe so calamitous that it had even obscured the sun. Viracocha apparently then set about civilizing the Americas, bringing back the sun and its light. So, in the context of this ancient myth, as with the myths of the life of Osiris, light and the wisdom of this great civilizer are very closely connected.

As we have noted, the idea that spiritual or psychological harmony is intrinsically connected with light is one of the most enduring in history. Read the book of Revelation, for example, one of the most powerful of scriptures, and you will see musical structures and references to light leaping out from every one of its twenty-two chapters. Similarly, the Passion itself, an eight-day event that culminated in Christ transcending onto the greater scale above, is in essence virtually identical to the myth surrounding the life, death, and resurrection of Osiris. Significantly, the "Passion" of Osiris, which was first enacted publicly by the Egyptian priesthood in Abydos during the Twelfth Dynasty, consisted of eight consecutive performances.

According to Christian tradition, on the eighth day of the Passion,

Yahweh !

Jesus floated upward on a cloud to heaven. The cloud is not without symbolic meaning, because it was through a cloud that God is said to have spoken to Noah and later to Moses of his covenant, symbolized by the rainbow, the spectrum of light, the "seven spirits" of the God of Revelation.

Many other examples of religious doctrines embracing one or another aspect of musical theory have been discussed at length in *The Infinite Harmony*. But the important point about all of these belief systems is that through this common principle of a harmonious development up through the earthly scale of our origins and on to a "heavenly" scale, these highly potent teachings, anachronistic though they may appear, are even today continually drawing the attention of billions of devotees upward. It's as if all of mankind's higher thoughts and aspirations are inevitably light-bound, heading—quite literally—for the sun and the stars.

The fact that these belief systems are still forces to be reckoned with suggests to me that "religious" concepts and precepts, being hermetic, or harmonious in every way, are quite naturally fixed in the memories of whole populations, not merely for a few years or so but for centuries and millennia. This, I believe, is what Gurdjieff would call real or "objective" art.

In the ordinary sense, of course, we cannot touch, weigh, or measure religious concepts and symbols: they are "metaphysical." But they exist, in one form or another, in all our minds. We noted earlier how a number of investigators have independently suggested that people's thoughts, if they are forceful enough, may be as real as the ground on which we stand. Bohm and Gurdjieff, for example, each believed that consciousness is actually a rarefied, currently unmeasurable form of matter. If this were so, then, theoretically, this ephemeral form of materiality known as a "concept," once created, would have the potential to exist independently of the individual mind that conceived it. This would, perhaps, explain why the extraordinarily resilient ideas and concepts of the Egyptians or the Greeks, or of individuals like

Muhammad or Isaac Newton, Moses or Einstein, are still around for all to "see," because, being in essence psychologically harmonious, they are highly resonant "things" in their own right. The Egyptians, I believe, regarded concepts in precisely this way, as a qualitative form of resonance operating according to musical laws. Viewed as such, religious, philosophical, and indeed seminal scientific concepts and ideas can be envisaged as metaphysical "notes" in the unfinished symphony of mankind's evolution.

Go at any time into a cathedral, a mosque, a synagogue, or a science faculty and you will witness the direct effect of these evolutionary metaphysical "notes," these concepts, on all those within. All of these human activities, all of the emotions and thoughts involved, are "light-bound," the residual product of human evolution, slowly gathering and increasing in rarefaction toward a condition of optimum psychological resonance. Generally it's a slow process, but this is because most social animals live not by the higher ideals and precepts of a Christ or a Buddha, but by Darwinian principles, through which changes, or beneficial mutations, happen only very rarely. Fortunately, perhaps, we don't all have to live like apes, because a way out of the Darwinian mode has already been charted by our early ancestors, and the sun, the stars, and even the galaxies themselves are all stations en route.

All this, of course, has staggering implications, because it suggests that the ancient Egyptians were in certain respects psychologically more advanced than we are today. Through some kind of practical application of the Hermetic Code, a key feature of which was the "pyramid ritual," these people managed to "enlighten" themselves, to climb up onto the higher plane of light, and go down into the quantum universe of the photon quantum. The hermetic phrase "As above, so below" expresses this concept perfectly.

The plane of light—the physicists' "quantum field"—permeates the whole of the material world existing in time. And, as we have noted, the omnipresent photon is the "force-carrier" of all quantum processes, the intermediary between all electromagnetic interactions. So when matter changes, say, by transmuting under intense heat and pressure, as when carbon-based compounds turn into diamond, or by decomposing, as in the oxidation of metal or the weathering of stone, photons are continually being absorbed or radiated by electrons in kaleidoscopes of highly resonant particle/wave activity. So if, as Yogananda asserted, it is indeed possible for the disciplined mind to tune in to the optimum harmonic frequencies at which photons resonate, and thereby enter the timeless, spaceless heaven of the ancients, then we are considering here access to a higher scale or plane of existence that in fact reaches right into the very heart of the electron, one of the basic constituents of all matter.

We have already seen that, according to many ancient myths, the Egyptians and their Native American counterparts used "music," or some form of sound technology, as an aid in their construction techniques, particularly in respect of the movement of heavy blocks. We also noted Andrew Collins's investigations into sonic technology in his book *Gods of Eden,* in which he describes eyewitness accounts of travelers to Tibet in the 1930s who saw the apparent levitation of stone blocks actuated by priests using numerous customized instruments. According to one witness, the mysterious Dr. Jarl, the use of musical or sound instruments by the priests in these demonstrations appeared to have been "accompanied" by silent "players" in the drama, namely the two hundred or so monks standing in rows eight to ten deep behind the musicians themselves, whom Jarl suggested might have been contributing toward the procedure by applying some form of coordinated psychokinetic force to influence the outcome of the event.

We are talking here of something very similar to what Colin Wilson has called the "group-mind" situation, the notion expressed way back before the time of Plato in the form of the Greek concept of a state of homonoia. Possibly, therefore, the instruments used by these Tibetan priests were effective primarily because they had been devised and subsequently activated by a highly trained collective of psychologically harmonious individuals, enlightened people whose minds were already "in tune" with what Schwaller de Lubicz described as "all the rhythms and harmonies of the energies in the universe."

The highest and most resonant of the "energies" alluded to here is, of course, light itself. As we have noted, the velocity of light is an absolute physical law. It is also the key to the timeless, nonlocal plane of light, the fifth dimension, defined mathematically by physicists as a sphere of existence in which there is no time and space. This is the dimension that I believe is described in myth by the Egyptians as the Kingdom of Osiris or the Duat, which refers both to the starry world above (the higher plane of light) and the mysterious underworld below (the nonlocal, quantum field of the subatomic particle). And Osiris, of course, who had the ability to perceive both of these domains simultaneously, was head of the musical pantheon of eight gods, whose principal monument —the Great Pyramid—was primarily associated with the phenomenon of light, which is itself a musically structured phenomenon. Thus we have a whole series of very close connections between the "builder gods," music and light. Add to this equation a correspondingly high level of consciousness (which we know existed at that time, because "it" conceived of the Hermetic Code) and we may well have all the ingredients necessary for the optimization of any activity, whether it be building a pyramid or simply sweeping a temple floor. Of course, identifying all the ingredients is one thing, but understanding how to combine them, and in what measure, is quite another. It is this distinction, one suspects, that marks the real difference between the ancients' intuitive right-brain knowledge system and our own fragmented left-brain method, which is the difference between feeling something in one's bones and merely knowing certain associated "facts."

Therefore any number of us today might go out into a meadow en masse and try to mimic the exercise described by Jarl, detail by detail, with disappointing results, because we would be merely aping, lacking the experience derived from long periods of disciplined, serious work involving systematic exercises in meditation, in "stilling" and subsequently developing the powers of the mind. Heavy blocks of stone would very likely remain just that, solid lumps of matter locked in a universal and inviolable gravitational field, in which everything is permanently endowed with a tangible property known as "mass." But then we are not trained ascetics; we are predominantly secular, with secular demands made upon us, and we have neither the time nor the inclination to spend years acting out the "pyramid ritual" in a disciplined way. Maybe if we had, like Yogananda, for example, or Jarl's Tibetan hosts, we might see "things" in an entirely different light.

The possible methods of manufacture and construction employed by the stonemasons of ancient Egypt are currently the subject of much heated debate. By and large, everyone seems to be genuinely baffled.

Currently in focus are a number of controversial suggestions as to the engineering techniques used by these "primitive" construction teams, such as, for example, Christopher Dunn's ideas about sonic/ultrasonic stone carving and drilling as outlined in his book *The Giza Power Plant*. The latest data, both the pros and cons of Dunn's ideas, were for a time posted regularly online, so we need not dwell on them here: suffice it to say that the question of machining techniques in the distant past is far from resolved. Andrew Collins has also contributed to the debate with his investigations into the ancients' sonic technology and the possible use of "sonic platforms" in the raising and transportation of their megalithic blocks. The description of the stone-raising techniques of the Tibetans by Dr. Jarl further implies the possible involvement of psychokinesis in the procedure: use of the homonoic technique.

Inevitably orthodox scholars will reject such notions outright. The general consensus is, of course, that the ancient stonemasons and builders used "primitive" methods only. Presumably this even applies to the four gargantuan monoliths incorporated several courses up in the retaining wall of the Temple of Baalbek. These blocks, remember, whose combined mass is estimated to be a staggering three thousand tons, were cut, perfectly shaped, and then transported to Baalbek from a quarry several hundred meters distant. While it may be difficult for us to imagine a scenario in which these giant stones were made to resonate in such a way as to make them temporarily weightless, the proposition is no more fantastic than the conventional position, which holds that these giant blocks were transported this distance and then raised using only ropes, rollers, and wooden levers. Indeed, of the two scenarios, the first seems more likely: it does at least fit the bill, whereas the "primitive" answer patently does not. And then we have the "musical" myths, of course, which speak of "builder gods" who could make blocks of stone float through the air simply by whistling or playing sound instruments. Significantly, there are no myths about "magic" ropes or "charmed" levers. There is only music—music and a "union of minds."

So what these myths tell us, in fact, is that the ancient builder gods had somehow discovered a way to effect a powerful psychic interaction between mind and the elemental world of matter. Sound may have played an important part in the procedure, but consciousness itself, through some kind of union with light, would have been the prime mover.

We have noted that Jahn and Dunne's experimental research has consistently produced statistical data indicating that most people possess a weak psychokinetic ability. They believe that psychokinesis is possible because consciousness itself is a kind of particle/wave phenomenon, with its wave mode, like all waves, capable of producing effects at a distance. Like the ancient Greeks and the yogis of the East, they do not see these psychokinetic effects as one-way processes, but rather as complementary exchanges of "resonance" between the thinker and the object.

Not too long ago, ideas like this would have been summarily laughed out of court, but when one hears today's scientists talking of particle consciousness, of "mind-like" electrons and "telepathic" photons, it begins to look as if anything is possible. Furthermore, the suggestion that the mind can somehow generate sufficient force to collapse the wave-packets of quantum systems outside the brain is in no way ruled out by these latest observations. In fact, when considering the nonlocal "action at a distance" between correlated photons, one might reasonably say: if fundamental particles can do it, then the human brain itself, an almost supernaturally well-coordinated mass of trillions upon trillions of highly active wave/particles, can perhaps do it infinitely better.

Interestingly, scientists are now trying to understand all physical phenomena not as isolated entities, but as integral parts of a single but much wider picture of reality, one that, significantly, also includes the mind of the observer. Particles, we are told, manifest as such only when certain of their properties are "seen," when they have been detected by an investigator—usually through annihilation of the particle and analysis of the debris. Without the participation of an observer, it seems, "particles" as such don't exist; they remain, as Bohm says, "enfolded," in a wavelike state of limbo. So the two aspects of quantum reality—the observer and the observed—are now seen as integral functions of the same phenomenon. Obviously, introducing this psychological element into scientific investigation is an important development, because it is leading scientists on to question the nature of their own consciousness. Possibly, therefore, the next generation of physicists will ultimately become "yogis" in their own right, able to experience for themselves the fundamental laws they have for so long been trying to formulate. Certainly, if there is a continuation of present trends by which the dividing line between scientific thought and metaphysics becomes ever fainter, we would do well to watch this space.

We will be returning later to the question of some kind of psychic element being involved in the construction of the ancient buildings of Egypt and the Americas. But in the case of the Giza necropolis in particular, another important question is, of course, why? Why did these early masons take the trouble to build on what is, even by modern standards, an incredibly vast scale?

The answer, it seems to me, is that they were totally and selflessly committed to the task of transmitting the essence of their ideas about light, music, and consciousness out into the exoteric world, and they obviously realized that the most effective and enduring way to do this was by "writing" all this data not on perishable parchment but in stone. Thoth himself, the originator of the Hermetic Code, whose followers designed and constructed the Giza site, was known in ancient times as the "scribe of the gods," a writer no less. We have all heard it said how much mightier is the pen than the sword. Nowhere is this adage more applicable than at Giza, where the "scribes" used quills the size of pneumatic drills and wrote in gigantic, three-dimensional "letters" across acres of bedrock.

The magnificent architecture of these Masonic scribes seems astonishing to us today. But the old and weathered physical remains of this great builder culture are really only a tiny part of the greater edifices constructed by these remarkable individuals.

So the Great Pyramid, the most impressive monument to light ever created on Earth, massive and imposing as it is, is really no more than a foundation stone upon which has been constructed another, infinitely vaster, metaphysical structure, a creation of sorts, whose indeterminate dimensions are even to this day expanding ever outward and upward. I am referring here, of course, to the ongoing evolution of human consciousness, which began its present stage of development at the time the Great Pyramid was designed, and which has ever since been guided subconsciously by the all-embracing hermetic principles embodied within it.

The Hermetic Code, therefore, is an evolutionary code. It describes exactly how DNA and the genetic code operate in the creation of greater organic structures, and for the last five thousand years it has been the basis of every major religion on Earth, movements designed specifically to facilitate the continuing evolution of human consciousness into higher "scales" of existence. Everything else we might surmise about the knowledge of the originators of this code is secondary to this fundamental concept. The Egyptians were brilliant architects, master craftsmen, highly accomplished astronomers possessed of the details of geodetics and precession, but they were first and foremost evolutionists, people who fully understood the underlying harmonies and rhythms inherent in the creative processes of nature and who conducted the whole of their lives in accordance with them.

So the "Egyptian way," the art of the alchemist, was the path of "creative evolution," an organic system of development and spiritual growth that fully complemented the evolutionary forces of nature. In the following chapter, in which we look at some of the ideas of the modern evolutionist, we shall see that this remarkable system is as meaningful to us today as it was to the ancient Egyptians.

6

Live Music

We have seen how the modern scientific description of physical reality, in many ways echoing the voices of thinkers long passed, encompasses the idea that everything, even something as seemingly inert as a lump of rock, is in some inexpressible way alive. David Bohm summed up this latest view by describing the electron, the most substantial component of matter, as a mindlike entity. The "choreographed" movements of electrons in plasmas and metals reveal that there are hidden orders implicit in the greater quantum field, where everything appears to be interconnected and potentially "aware" of the presence of everything else. So, in the opinion of many modern physicists, nothing in this universe is truly dead: everything resonates, communicates, radiates, and absorbs. And, as we have seen, there is a clear musical pattern to it all.

Now, when we consider entities that are organic and alive in the sense in which we normally understand the term, we find once again that there are clear musical symmetries evident in their underlying structures. Specifically, the biomolecular world, as I explained in the Introduction of this book, is an endlessly unfolding symphony of "live" music, of genetic harmonies and interpenetrating organic scales. Remember that the Hermetic Code, which is in essence a musical system, describes in precise detail how the genetic code works.

It is interesting that this altogether remarkable fact has not yet been acknowledged by the scientific establishment, even though it was first made public back in 1994.[1] This is, however, not just any old "fact" we are asking these scholars to consider. It is an all-pervading, fundamental truth, one that carries with it the unavoidable and, perhaps, unpalatable, conclusion: that the latest picture of the organic world being described by modern biologists, like the physicist's description of the underlying structure of matter, is basically just a cover version of the original canon first revealed by the original hermeticists. What we are saying here, in effect, is that the historian's mysterious "hunter-gatherers," who populated the more temperate regions of the Earth at the dawn of recorded history, were apparently more in tune with life's creative processes than any scientist alive today. We have proof of this. We have the Hermetic Code, the code of life itself, described by a simple formula that embodies an idea in circulation for untold millennia, an idea that is, in fact, so profoundly relevant to us all that it will never fade away.

By way of finding further proof in support of the above heresy, we must now make another "musical" journey through time. It's a very long road indeed that you are about to travel, for we shall be going way back, beyond the eras of the Greeks and the Egyptians, back to the music of the Neanderthal race and further still, beyond even the time of the dinosaurs—right back, in fact, to point zero, to the very first evolutionary "note," the first primordial spark of life on Earth. As we shall see, the Hermetic Code has been in evidence practically from the very beginning.

The present, most widely accepted account of our origins is of course the Darwinian theory of evolution, which asserts that you and I and the consciousness we are endowed with happened to have evolved here on Earth purely as a result of blind, accidental physical and biochemical processes. This basic concept, the random evolutionary development of life, has in the last few years become virtually a scientific dogma. When we compare ancient and modern ideas on evolution, however, we find that the latter is but a pale imitation of its predecessor. This is not to say, of course, that Darwinism is not a valid theory of evolution, only that it doesn't go far enough. So let's see in what respects this great theory falls short of the original described by the priests of Old Kingdom Egypt. We can begin with the experts.

In a recent book, *River out of Eden,* Professor Richard Dawkins, author of several influential books on evolutionary theory, attempts to explain the whole phenomenon of life in terms of Darwinian principles. This is a theory that he fully endorses and which, he says, displays "a sinewy elegance, a poetic beauty that outclasses even the most haunting of the world's origin myths."[2]

Already, it seems, I am inescapably and completely at odds with one of the world's foremost proponents of current evolutionary theory, who clearly believes that the modern scientific interpretation of our creation is superior in every way to the ancients' description of mankind's origins. So let's see.

Dawkins's lucid account of how life evolved on this planet constitutes an impressive argument in favor of the theory of natural selection. According to this view, all living creatures are indirectly descended from a single, primitive ancestral species, which evolved and diverged into new species over billions of years through random copying errors in DNA replication. Many of these mutations of single genes will have had deleterious effects on the functions of the host organism and actually reduced its chances of survival in the world. Very occasionally, however, a gene-copying error resulted in a change in the organism's functioning that happened to be beneficial in life, improving its performance in some way and so increasing both its chances of survival and its ability to produce equally successful descendants.

Clearly Dawkins has little time for the Creationists' arguments against the apparently random, ungodlike nature of gene mutation and natural selection. To be fair, Creationism (as interpreted by modern theologians) is not a theory at all; it is simply a blind faith, whose advocates have haphazardly concocted a rather flimsy file of uninformed criticism of Darwinism, none of which provides convincing evidence for their ideas.

There is a paradox here. While Creationists have been busy nit-picking at scientific theory, they have all along been in possession of the complete evolutionary picture. The whole story is there, as we shall see, in their scriptures. And this account includes a detailed description of the newly discovered biomolecular world, albeit only as part of a much broader and more comprehensive theory of evolution.

The commonest criticism of Creationists relates to one of the main problems of Darwinian theory, the difficulty it has in explaining how such an intricate organ as, say, the human eye was formed. What, for example, did the intermediate, developmental forms of such a sophisticated organ look like? What kinds of beneficial evolutionary functions did these earlier, rudimentary conglomerates of cell tissue facilitate before they evolved into a state that actually bestowed upon the host organism the ability to even recognize the difference between light and darkness, let alone "see"? Darwin himself said that he could never imagine the eye, with all its structural complexities, as having evolved through random variation and selection alone. Dawkins disagrees. In fact, answering this rather difficult question is, he assures us, "a doddle." Half an eye, he argues, is 100 percent better than no eye at all, 1 percent better than 49 percent of an eye, 1 percent worse than 51 percent—and so on, providing an evolutionary scenario that suggests a gradual, hit-and-miss process of development.

Now this might seem at first to be a fairly reasonable line of deduction. But then, what about a measly 1 percent of an eye? One percent, 2, even 3 percent would be so far removed from an organ that sees, or even one only half-formed, that it is difficult to envisage what kind of survival advantages these first crude mutations would have given to the evolving species. Would 2 percent of an eye, for example, give a creature even the slightest hint that one of its natural predators was about to pounce from a short distance away?

There are other complex evolutionary developments that are difficult to explain solely in terms of Darwinian theory—for example, the evolution of fins and wings. We are told that fins evolved into hands and that arms evolved into wings. But of what use were these different functional structures during the in-between stages of development? What advantages over others would a species existing through the proposed transitional stages have with an appendage that was neither arm nor wing, but a bit of both? If and when traces of the existence of such peculiar creatures are ever found in the fragmentary fossil record, then perhaps a little light might be shed on the problem. To date, none have been identified, and one can only speculate on the reason. The Harvard paleontologist Stephen Jay Gould has suggested that such transitions may have been discontinuous, rapid changes, invoked by single crucial genes that somehow managed to cross over concurring pathways of development and so affect the structure of the whole organism in a much more radical way.

The only creature remotely resembling an "intermediate" that has been identified in the fossil record is the archaeopteryx, a raven-sized creature with a reptilian skull and birdlike wings. Its wings were fully formed, however, a fact that has prompted many biologists to classify the archaeopteryx not as a true intermediate but as a completely developed bird. Indeed, it is difficult to imagine how any true intermediate with a confusing complement of half-formed wings could have survived at all. Should it attempt to fly away from danger, or jump? In that almost comical split second of indecision it would probably have been summarily dispatched by some ravenous carnivore. This might account for the fact that no half-formed wings have ever been identified in the fossil record. Presumably such unfortunate creatures would have been hopelessly ill-equipped to survive long enough in a ruthless, *lex talionis* world to produce offspring in sufficiently large numbers.

Returning to the question of the eye for a moment, if Darwin's theory isn't quite the whole story, could there possibly be other factors involved in the formation of such a complex organ? That is, could there be other agencies involved in the evolutionary process that might be to some extent responsible for the eye's amazingly intricate development?

I would suggest that there is at least one agency that might be worth considering here, a phenomenon that is intrinsically connected to the eye like no other. Furthermore, it is all-powerful and omnipresent, and occupied a uniquely important position in the belief system of the first true "Creationists." This, of course, is light, the Holy Ghost, the "Rainbow Covenant," the agent of all visual sensation. Obviously if there were no light in the first place, an organ built to perceive it would never have evolved. We know that primitive sea creatures living at the bottom of the deepest oceans, where no light ever reaches them, are devoid of normal photon receptors. So the mere presence of sufficient light is enough to induce the development of organs capable of assimilating it. Remember also that light itself, as well as being the agency by which objects are made visible, is also, as the ancient Egyptians were aware, a musical phenomenon. With its seven fundamental frequencies (the spectrum) and its three distinct "primary" frequencies, it is a perfect electromagnetic model of the "triple octave," of the Hermetic Code, which describes the evolutionary development of all organic processes, including, of course, the process of the development of the eye.

The "river" in the title of Dawkins's book refers to the stream of digital information flowing through time in the form of DNA, branching and forking on its way, and giving rise to new species in the process. When one species mutates off a daughter species, the river of genes forks into two, and if the two tributaries diverge for long enough, perhaps through the external influence of geographical variables, the two species then develop quite distinct and different characteristics. At one stage, apparently, in one of these branches, Mitochondrial Eve was born. This is the name given to the most recent common ancestor of all modern humans, a member of the species *Homo sapien sapiens* who probably lived in Africa between 100,000 and 250,000 years ago.[3]

The name of this "hybrid from Eden" is derived from the term *mitochondria,* which are vital, energy-producing "particles" existing by the thousands in all our cells. They help to convert energy from food molecules and then store it for distribution as and when required. The significant point about mitochondria is that they have their own DNA. Unlike the main DNA housed in the nucleus of the cell, which becomes almost totally scrambled in every new generation every time a sex cell is made, mitochondrial DNA is passed down relatively unchanged through the female line only. It is therefore a very useful tool for long-term genealogists, who can use it for dating common ancestors within species. This is how Mitochondrial Eve has been identified and dated.

Dawkins is particularly fond of Mitochondrial Eve, and he contrasts her, as a scientific hypothesis, with the Eve of Eden. He believes his "scientific truth" to be of greater interest and, I quote, "more poetically moving" than the original myth.[4]

Presumably, at the time of writing this, Dawkins was unaware of the existence of the Hermetic Code, of the fact that it is identical in every way to the genetic code, and also of the fact that all major creation myths are in effect variations of the same, original theme. This, as we have seen, is not simply the spurious product of primitive superstition and folklore, but a genuine scientific theory, presented symbolically in the clearest of terms: it states simply that all creative processes are the product of forces described by the two fundamental laws of nature embodied in the *pi* convention, the law of three and the law of octaves. Clearly, it is only in this context—that is, from the hermetic perspective—that the original story of the Eve of Eden can be truly understood. In fact, as I explained in *The Infinite Harmony,* much of the Hebrew Scriptures, from the creation of the world and the story of Eden, through to the charmed lives of Noah, Joshua, Moses, David, and Solomon, contains innumerable symbolic musical references to the Hermetic Code, to the theory of transcendental evolution. And, remember, the creation myth of Genesis is just one example, one old "fairy tale" from the Middle East. There are many more, of course—Chinese, Vedic, Zoroastrian, Christian, and so on—and they tell exactly the same tale, providing a truly accurate description of the natural processes of creation. Think of the I Ching, and how perfectly its overall format corresponds to the structure and symmetry of that remarkable biochemical code used by DNA. Are we to believe, then, that the emergence of these identical symmetries are both merely the product of "accidental mutations"? I really don't think so. One of these symmetries—the genetic code—just might have originated by chance, though personally I doubt it, given the fact that the physical universe itself is structured along the very same musical symmetries. But then, when we see exactly the same symmetries being repeated yet again in a higher scale of evolution, that is, in the mind of man, it begins to look as if it might be people like Dawkins, and not the ancient mythmaker, who is unwittingly purveying the fairy tale.

This repetition of identical symmetries in scales both "above" and "below" is important, because it suggests that there is an underlying unity and sense of purpose in all life. This purpose, quite clearly, is to evolve, and to do so musically, transcendentally—just as DNA has done over the past four thousand million years. You see, the DNA symmetry is not only "resonating" in our blood and in our bones, the same symmetry is also active in our minds, enlivening them, vitalizing them, coaxing them ever upward. Possibly this is why I am sitting here now, writing all about this symmetry, because, deep down, somewhere inside me, DNA is telling me to.

This now leads on to an important question: who or what told DNA how to behave? Dawkins's answer is unequivocal: no one, nothing; it found its way quite by accident, stumbling blindly through geological time, through endless cosmic and geophysical cataclysms, ice ages, or whatever.

It is true that much of DNA's evolution on Earth has been erratic and at times very chancey, with many of nature's experiments (like, say, dinosaurs) going drastically wrong. But remember, underlying all this apparent random, selective evolution is the symmetry of DNA and the genetic code, a symmetry that, as we have seen, is actually controlled by the forces described by nature's two fundamental laws. So this is in no way simply the product of chance. DNA's distinctive form and method of evolution is an inevitable consequence of these natural forces: it was preordained by nature itself.

It seems to me that the main problem with Dawkins's position is that there is little or no music in it, no allowance made for the evident hermetic symmetry of the biomolecular world. This is where ideas old and new really diverge, because the mythmakers, the originators of the Hermetic Code, of *pi* and the "triple octave," knew all about this music and about the laws and forces that conduct it.

As we have seen, the Hermetic Code is much more than a mathematical tool. It is a universal blueprint for all evolutionary or creative development, and its distinctive inner symmetry is to be found in the biomolecular and physical structures of all forms of life. We see this not only in the sixty-four-word/twenty-two-note amino acid "scale of resonance" but also, for example, in the overall physiology of human beings, with their three nerve complexes responsible for sensation, emotion, and perception, and their eight sets of endocrine transformers, the glands responsible for secreting into the bloodstream all the drugs and hormones necessary for enabling reaction to external stimuli. Thus, all human beings, and in fact all living things, are hermetically composed; they are all in their relative scales evolutionary "triple octaves" with the inherent potential to achieve a state of "optimum resonance."

The living cell uses these hermetic symmetries to sustain itself and to develop. Ultimately it attains the necessary condition of "optimum resonance" and so acquires the supernatural power to self-replicate. Through a systematic sequence of exponential growth patterns it then combines with its fellow cells to create a whole new world for itself, a massive, complex, conscious organism. Such an organism constitutes a higher dimension for the cell, a higher "scale of being."

The theory of transcendental evolution, the essence of which is contained in the now familiar phrase "As above, so below," asserts that, at the human level of existence, it is possible for individuals to emulate the living cell and to achieve a similar condition of "optimum resonance." Traditionally this unique condition of being has been most commonly acquired by following certain tried and tested "religious" codes of conduct—the idea being that such practices eventually endow the individual with special powers: to "self-replicate" in some way, to create whole new worlds, to penetrate up into an infinitely higher scale of existence, the scale which ultimately became known as heaven.

So *pi* itself, the blueprint for the evolution of all life, is also the blueprint for the evolution of consciousness. It represents an exact scientific description of the optimum metaphysical frequency, an "immaculate" psychological wavelength accessible to us all, through which mankind can ultimately break free from the ponderous mode of evolution characteristic of the "naked ape."

Thus, the Hermetic Code describes the fundamental organic matrix upon which we have all been constructed. Whatever else we might care to think of ourselves, we are fundamentally walking "trinities," triple octaves of resonance, comprising our sensations, emotions, and perceptions. And according to the originators of the theory of transcendental evolution, the three nerve complexes controlling these vital functions can be systematically developed up to a point where they all "resonate at optimum potential" and so acquire the power to transcend on to a higher scale of existence. Clearly such a condition of being is a far cry from our present evolutionary state. As Gurdjieff and Schwaller both said, somewhere along the line our ancestors lost the plot and slipped back into a rudimentary Darwinian mode. I believe the mythmakers and metaphysicians of ancient days foresaw this decline, and that this is why they went to such great lengths to project the Hermetic Code out into the greater sphere of humankind's collective consciousness. They knew that this sacred concept would lie dormant, like a seed in the soil of the subconscious mind, but that sooner or later this seed would germinate, take root and grow, and ultimately flower and bear fruit. This description is not intended to be taken as a metaphor. The process outlined, as we shall see subsequently, is organic from beginning to end. And so it is that today, now that our level of comprehension has reached, as it were, the necessary "pitch," we are witnessing a worldwide "Egyptian renaissance," what we might call a new flowering of awareness and appreciation of the great wisdom and remarkable abilities of the metaphysicians of ancient times. Accordingly the Hermetic Code itself has surfaced once again, and its symmetries have been recognized, not only in all the major scriptures and in the dimensions of ancient pyramids all over the world, but also in our blood, in the white ray of physics, and in the underlying structure of the entire physical universe. In truth, now that we have eyes to see, we find that these symmetries are everywhere.

I stated above that I believe that the growth and development of consciousness is an organic process. Logically it has to be, because the Hermetic Code and the genetic code are fundamentally one and the same system. And, in fact, this organic correlation is further compounded in the ideas of Pythagoras, the main proponent of hermetic theory in ancient Greece.

The Pythagoreans themselves left no written records. The "Golden Verses," whose authorship is generally attributed to their founder, may be authentic, but they are scanty and fragmented and contain no hermetic data as such. What the Pythagoreans did leave for posterity, however, was a comprehensive array of esoteric symbols: numerical, geometrical, and, of course, musical. The language of symbolism was their way of recording and transmitting their ideas, and when we examine the most "sacred" of these symbols, we often find that they possess a number of distinct but related facets. Of course, *pi* is the prime example: it conveys mathematical, geometrical, musical, mystical, and even scientific truths, all neatly condensed into a single, imperishable sign. Another significant symbol to which Pythagoras attached great importance was the sacred "Tetrad." This was expressed by placing ten pebbles on the ground like so:

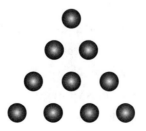

Like *pi*, this is in essence a musical symbol, another symbolic expression of the Hermetic Code and was referred to in Pythagorean schools as the model of the gods. Also described as the source of nature, the Tetrad was seen as the fundamental matrix upon which to create the perfect individual, a "model" of the gods.

When we look at the configuration of the ten pebbles, we see that they are laid out in a 4–3–2–1 format, the whole depicting an evolutionary process developing from bottom to top. It so happens that this distinctive pattern of development describes perfectly the four distinct stages in the synthesis of amino acids, the very building blocks of life.

It will help to remind readers here that the DNA molecule works with the four chemical components, or "bases," of the genetic code. These are used to construct small molecules known as RNA triplet codons, comprising three bases each, which then serve as templates for the production of amino acids. The amino acids are then, in turn, assembled into the much more complex protein chains.

The Hermetic Code, as we know, is primarily an expression of the law of triple creation, which holds that everything is composed of trinities within trinities. This means that the three individual octaves embodied within *pi* are in themselves triple octaves, making nine octaves in total, or sixty-four "notes"—precisely the number of RNA codon combinations.

Look at the musical structure of the formula *pi* when set out in diagrammatic form:

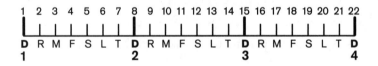

As we see, the Hermetic Code, just like DNA and the genetic code, is constructed upon four fundamental "base notes." These are represented in the base line of the Tetrad, which we have noted consists of four pebbles. If we now follow the successive stages in the synthesis of the amino acid, we see that the Tetrad describes this process exactly. Thus, from the four nitrogenous bases, DNA programs three of them at any given time (RNA triplet codons) to produce two distinct properties (amino acids are both acidic and alkaline) of one biochemical unit in a higher, more complex scale of existence, that is, one of the twenty-two evolutionary signals at the amino acid/protein scale of development.

There is one particularly detailed description of the Tetrad, written more than 1,500 years after the Pythagorean era, which is especially interesting, because it demonstrates even more convincingly how perfectly in tune these hermetic initiates were with the vital process of the creation of life, even to the extent of understanding the dual nature of what was being conceived in this way. The text in question, dealt with in some detail in *The Infinite Harmony,* is a tenth-century Egyptian commentary on the Koran, the Tafsir ol-Jalalayn, which gives an account of Muhammad's famous night journey to heaven. It begins with him riding on the back of a quadruped, sees him prostrating himself three times in prayer, after which he is offered by the angel Gabriel two cups to drink from, one of wine and one of milk, and finally, after wisely choosing the milk, escorted in triumph to the first of the "seven heavens."

Clearly this commentary on the Prophet's spiritual awakening is describing a hermetic process, the organic "flowering" of Muhammad's consciousness. It is a clear description of the pattern of evolutionary development of the Tetrad, which, as we have noted, depicts with absolute precision the way in which Watson and Crick's famous double helix is evolving through time. The two symbols of wine and milk are especially significant, because they describe perfectly the dual, acid/alkali properties of the amino acid.

We have now established that hermetic is genetic. This means, in effect, that certain fundamental aspects of consciousness—ideas, concepts, revelations, and so forth—are metaphysical genes and are produced in exactly the same way as are amino-acid chains.

Interestingly, this particular idea (or "gene") was first tentatively put forward by Richard Dawkins himself in his highly acclaimed first book, *The Selfish Gene.* Dawkins defines this new kind of replicator as a "unit of cultural transmission," or of "imitation," one still in its infancy but which "already is achieving evolutionary change at a rate that leaves the old gene panting far behind."[5] His chosen name for this metaphysical phenomenon—*meme,* from the Greek root *mimeme* (hence, mime, mimic, copy)—has since been incorporated into the main corpus of our language and has spawned a whole new embryonic science: memetics.

Examples of memes are numerous. Dawkins cites things like tunes, ideas, catch phrases, fashions in clothing, ways of making pots and of building arches. In fact, the list could be endless, because it would encompass anything invented by any given individual, good or bad, positive or negative, that is subsequently copied by another or others. This would include concepts as diverse and far removed from one another as fascism, belief in god, a scientific theory, the wheel, a literary genre, computer hacking, praying, killing, or whatever. Memes are likened to viruses, willy-nilly infecting or contaminating brains as they hop from one to another like subquantum fleas, apparently going nowhere in particular. This means that they are not seen as conscious agents but, like genes, as agents of "blind natural selection." So memes, or ideas, another form of replicator, are everywhere, floating awkwardly around in the primeval soup—the soup of human culture—endlessly parasitising our brains at every twist and turn. Even the great religions, philosophical movements, and the concepts of the most influential figures in human history can be classed simply as parasitic "meme complexes," all swimming around in a chaotic meme pool.

Memes are seen as agents of blind natural selection, but the evident longevity of the most enduring creations and inventions of man have to be admitted. As Dawkins explains, an individual's genetic input to the ancestral line is halved as each generation comes forth, so it doesn't take long for it to reach negligible proportions. The memetic influences of the more notable figures in history—like Socrates, Leonardo, Copernicus, and Marconi—are still going strong.[6]

Although he makes an oblique reference to the power of the organized church, whose architecture, rituals, music, art, and written tradition he describes as "a co-adapted stable set of mutually-assisting memes,"[7] Dawkins sees nothing special in religion per se. Indeed, he is well known for being particularly vociferous, and perhaps with good reason, in his condemnation of the Christian fundamentalist elements in certain southern states of America, where attempts have been made to keep Darwin's theory off the educational curriculum. In some states it is still obligatory for schools to teach the naïve literal interpretation of the Biblical creation story alongside Darwin. Of course millions will agree that studying scripture of any religious persuasion can enrich and stimulate the mind, if only by freeing it to speculate on dimensions beyond the restrictive confines of mere survival. Possibly Dawkins himself would agree that written traditions are not all bad. But the Biblical version, along with all the other major doctrines referred to in this book, is a hermetic work, an all-encompassing treatise on universal symmetries, and to present it simply as a factual account of the origin of the world and of mankind not only misses the point entirely, but alienates virtually everyone of a scientific or rational bent, and in so doing actually hinders the flow of a vital stream of important knowledge. If such a state of affairs were to continue, the "soup of human culture" would simply stagnate.

Understandably, therefore, Dawkins tends to focus on peripherals, on the negative or contradictory aspects of formalized religion, such as the Christian idea of hellfire or the meme for religious blind faith, loosely referred to as the god meme. This, in my opinion, is where he misses out on the most important line of enquiry, one that must logically take us right back to the creators of these long-lived religious movements. In the case of the Christian phenomenon we need of course to focus primarily on the deeds and words of Jesus Christ, and to do this we have to strip away practically all of the subsequent trappings and subjective embellishments of the formalized hierarchy. If we do this we end up with the kernel of all Christian preaching—the original "meme complex"—expressed by Jesus himself through the unchanging symbolism of what is known today as the Passion. And what immediately becomes apparent when we look at the symbolism involved is that the eight days of Easter week, from Palm Sunday to the Sunday of the Resurrection, describe a musical event, an octave of activity, beginning on the note Do and ending, after seven intervals, on the very same note. So clearly the message encoded here is hermetic, a scientific description of the organic evolution of Christ's consciousness, which, like Muhammad's, successfully transcended on to the greater scale above.

Given that the Hermetic Code is the blueprint of all religious symbolism, we might say that it is one of the oldest memes floating around. More than that, like RNA or DNA in the gene pool, it is ubiquitous, and so deeply immersed in the human psyche that it looks to be unstoppable.

We now know that the Hermetic Code has many facets, expressing mathematical, musical, geometrical, scientific, and even cosmic truths, all ingeniously condensed into a single, imperishable sign: 22/7. One might say, then, that this code is a meme complex in its own right. But is it, as Dawkins assumes it must be, an unconscious entity, merely an agent of "blind natural selection"? The answer, it seems to me, is far from decided. In fact, it is difficult to see how the Hermetic Code could be defined as a blindly evolving, unconscious replicator. It was intentionally created and subsequently disseminated by the mind of some unknown genius for a specific and entirely selfless purpose—to enlighten lesser mortals on the ways of Nature and so facilitate the ongoing evolution of man's collective consciousness. And it is still with us today, this code, as vital and resilient as the day it was conceived, an undiminishing beacon of metaphysical light shining into every nook and cranny of human endeavor. Even in the modern Western world, where secular influences predominate and the so-called god meme is in recessive mode in the majority of human brains, the fundamental components of the Hermetic Code are everywhere, in the symmetries of the material world of elementary particles and biomolecules, in music, legend, folklore, fairy tales, customs, and so on, stretching right back to the dawn of civilization. Just think of the number 7, every other person's lucky number. So evidently appealing is this particular meme, if you won the national lottery with the number 7 and multiples thereof, you would almost certainly be sharing your diminished prize money with thousands of others. Of all numbers, this symbol, which is of course a symbol of the octave, is without doubt one of the most efficient "replicators" in the entire development of human thought.[8]

The point is, the Hermetic Code is not just any old brainchild. It is not a fashion, a craze, a catch-phrase, a political ideology, or a computer virus. It is an all-embracing concept, which not only describes the process of evolution but actually facilitates its ongoing development. In view of its uniqueness, it would be inappropriate to classify it along with the common meme. For this reason I choose to differentiate between memes and evolutionary concepts by referring to the latter simply as metaphysical genes. This would apply to all significant concepts, such as, for example, the highly potent symbolism of the Passion of Christ, the eightfold ways of the Buddha, Zoroaster, and Confucius, the structures of the I Ching and of the tarot pack, or the symmetries of creation described in Genesis, in the Koran, or in the cosmogeny of the builders of the Giza necropolis. These ancient concepts have survived for millennia and look set to last for many more to come. There are modern concepts, too, which might fall into the same category. For instance, Gurdjieff's system is likely to prevail because it is essentially a hermetic phenomenon, a faithful "copy" of the original. Arguably Darwin's rudimentary "conception" is in there too—along with the revelations of such as Watson and Crick or Einstein. Einstein's Theory of Special Relativity is a particularly potent "metaphysical gene" and has changed our view of the world forever. Not only does it focus on the constant properties of light, the Covenant of the Ages, but also, with its implications for the elasticity of space and time, it provides us with an intuitive glimpse into another dimension, the timeless, spaceless world of the great Egyptian sun-god, the plane of light.

So what is their attraction? What makes significant concepts such good replicators? The answer, in my view, is their hermetic content, their evolutionary bias. It is this kind of impetus that ordinary people like you and me somehow find irresistible, even if only on a subconscious level. But of course anything, any invention that serves to promote the evolution of consciousness, must by its very nature be based on "sound" principles and therefore be psychologically harmonious. Possibly, therefore, it is this integral harmony, or the quality of resonance intrinsic to the concept, that explains the secret of its longevity. This is to say that concepts possessing a high degree of resonance seem to be endowed with a special kind of metaphysical power, the power continually to "self-replicate," by harmonizing, blending in, with billions of human brains.

So the significant concept, the metaphysical gene, is an evolutionary impulse; it has an inherent tendency to rise above the surface level of the "soup of human culture" and, in musical or hermetic terms, is much more psychologically resonant than a mindless catchphrase or the latest gizmo.

In Dawkins's view, memes, or metaphysical genes, have nothing equivalent to chromosomes: "In general memes resemble the early self-replicating molecules, floating chaotically free in the primeval soup, rather than modern genes in their neatly paired, chromosomal regiments."[9]

Evolution is considered to have kicked in when the early replicators in the primeval soup began not merely to exist but to construct for themselves containers as vehicles for their continued existence. The replicators that won through were those that built "survival machines" for themselves to live in. So genes came first and the machines—chromosomes, cells, plants, animals, people—came afterward. We might say, then, that the first true survival machine was the DNA molecule, the chromosome. This would imply that, at the next scale of development, the scale of human consciousness, ideas and concepts will eventually build "survival machines" for themselves, possibly beginning with the metaphysical equivalent of the chromosome. Dawkins assumes that this hasn't happened yet, that memes, or metaphysical genes, whether in isolation or in loosely affiliated complexes, are still drifting clumsily through space and time like spores in the wind.

There is, however, an alternative way of looking at metaphysical genes, one that sees the "chromosomes" housing them—their "vehicles"—as already existing. To understand this we need first to look at DNA itself, the original chromosome. One of the chromosome's functions is to create proteins, which it does by sending out copies of parts of its internal structure in the form of triplet codons. These are ejected into the chemically rich liquid membrane of the cell to do their work, to code either for one of the twenty amino acids or for one of the start–stop signals. The amino acids are then assembled by other genetic components into molecular chains. A chain of amino acids makes a peptide, a chain of peptides makes a protein, and the protein codes for one or another of a multitude of chemical processes in the evolutionary development of the entire organic body, the whole "machine."

If we subsequently apply this model to the world of memes, we might say that the metaphysical chromosome is the brain itself, so that a momentary thought or idea would function as some kind of metaphysical amino acid, or a cluster of them, and a full-blown concept, with all its cognitive applications, would be the equivalent of a metaphysical gene. The product of such a "gene," continuing to evolve and self-replicate in millions of other human brains, we might regard as the metaphysical equivalent of the biologists' greater protein macromolecule.

If this is really how it is, and the evolutionary processes "above" are in essence the same as the processes "below," it would require an explanation as to how a three-dimensional organ like the brain could possibly be regarded as a scaled-up replica of the DNA double helix. This will be the subject of the next chapter.

Earlier I suggested that this "organic" process of transcendental evolution, of "journeying to heaven," involved attaining what I call an optimum degree of psychological resonance, one that would ultimately be "in tune" with the constant frequencies of light quanta. The implication of this is that light and consciousness are, in effect, opposite sides of the same metaphysical coin—the "coin" itself being the metaphysical equivalent of the amino acid, the transcendental product of an enlightened mind. Now this product—this impulse, idea, or concept—like the amino acid, is derived primarily from three basic components or "bases," namely sensation, emotion, and perception. The concept that arises from the harmonious interaction of these three metaphysical "bases," just like the acid/alkali structure of the amino acid, must also have a dual nature, which means that these products of consciousness should have a complementary opposite. This must be light itself. Light is, after all, the primary agent of all visual sensation. It is also the most resonant "octave" in existence. Most importantly, however, it is a perfect electromagnetic blueprint of the Hermetic Code.

The unraveling of the digital molecular structure of the DNA double helix has been hailed by Dawkins as the most revolutionary scientific discovery in history, an achievement that, he believes, should be honored as much as the works of Socrates, Plato, and Aristotle. Whether or not this will, in the event, turn out to be the case remains to be seen, but it seems to me to be a tall order. The Athenian adepts almost certainly achieved their present status primarily because they were all fully "in tune" with the tenets of hermetic theory. Socrates, for example, the first of the great trio and Plato's mentor, is known to have received personal instruction from the Pythagorean School on the island of Samos. In addition, all three were known to have respected and familiarized themselves with the "musical" traditions of the legendary Persian Magi.

The Greeks, of course, believed the original source of all this hermetic wisdom to have been the ibis-headed Egyptian god Thoth. Known to the Greeks as Hermes Trismegistus ("Thrice-greatest Hermes"), he later became associated with the Roman god Mercury, the messenger of the gods. Most of the extant literature relating to Hermes belongs to the post-Christian era, and scholars in general doubt his historical authenticity. But of course, someone, somewhere, revealed the Hermetic Code to the human race, so by whatever name this individual was then known, we can be reasonably sure that the originator of Revelation certainly did, at some remote period in prehistory, walk upon this earth.

We have already established that this remarkably astute observer understood fully how life is created. The symmetry of the Hermetic Code and the symmetry of the genetic code match too precisely for us to think otherwise. But in case there is still some doubt in the reader's mind, it is perhaps worth noting here another intriguing detail in the legends of Hermes/Thoth, one that, to me at least, is so fitting as to be more "poetic" even than Dawkins's Mitochondrial Eve. The rod or wand of Hermes/Mercury was known as the caduceus. He is depicted holding it in Botticelli's painting *La Primavera*. The wand itself, said to have had awesome magical properties, was surmounted with two wings and entwined by two serpents. It is a perfect double helix.

Significantly, this same design also appeared in ancient America, and in much the same context, that is, as a symbol adopted by a legendary man of high learning. In *Fingerprints of the Gods,* Graham Hancock describes a statue in Teotihuacán near Lake Titicaca in Bolivia, of the mythological civilizer of ancient South America, Viracocha. A bearded, Caucasian figure with features identical to the early depictions of the Egyptian god Osiris, Viracocha is wearing a ceremonial robe, on either side of which is engraved the image of a serpent, coiling from bottom to top. In another representation, Viracocha holds a thunderbolt in each hand. A thunderbolt, of course, is a stylized representation of a helix, and Viracocha is holding two of them. Further north, incidentally, Viracocha's Mexican counterpart, Quetzalcoatl, had as his symbol a plumed serpent—again very reminiscent of the plumed caduceus of Hermes.

Possibly the most impressive ancient representation of the double helix I have yet encountered is the pyramid Temple of Kukulkan (the name for Quetzalcoatl in the Mayan dialect) at Chichén Itzá, in northern Yucatán, Mexico. This is what Graham Hancock has to say about this remarkable structure:

> Its four stairways had 91 steps each. Taken together with the top platform, which counted as a further step, the total was 365. This gave the number of complete days in a solar year. In addition, the geometric design and orientation of the ancient structure had been calibrated with Swiss-watch precision to achieve an objective as dramatic as it was esoteric: on the spring and autumn equinoxes, regular as clockwork, triangular patterns of light and shadow combine to create the illusion of a giant serpent undulating on the northern staircase. On each occasion the illusion lasted for 3 hours and 22 minutes exactly.[10]

Two serpents coiling up to the sun, like starbound DNA. Esoteric, yes, but entirely comprehensible in the light of the theory of transcendental evolution.

Such remarkable similarities in ancient global symbolism may lead us reasonably to suppose, as Hancock, Bauval, West, Wilson, and others have suggested, that these legendary civilizers—Viracocha/Quetzalcoatl/Kukulkan, Osiris/Thoth/Hermes—if not one and the same individual, were an elite group from a forgotten race of people. Possibly they were survivors of the cataclysms of the last ice-age meltdown, who successfully disseminated their profound understanding of the laws governing creation and evolution across the entire planet at an unknown period of the distant past. But how distant?

Dawkins says that Mitochondrial, or African, Eve, was the first member of our own species, *Homo sapiens sapiens,* to produce descendants successfully. She is thought to have appeared possibly as recently as a hundred thousand years ago. Before her there were more primitive hominids, such as the Neanderthals (*Homo sapiens neanderthalensis*), *Homo erectus,* and its even more remote ancestors, *Homo habilis* and *Australopithicus,* the earliest having been dated to around four million years ago.

So a great deal has happened in the evolution of the hominid over the last hundred thousand years; for Eve, the mother of the modern human race, is thought to have been little more than a cunning savage, a successful fighter and breeder, an unwitting, cutthroat product of DNA's relentless bid for immortality.

This conventional picture of the primitive "savage" dominating the world stage around one hundred thousand years ago is by no means accepted by all scholars. We have noted already how Stan Gooch has shown that the predecessors of African Eve—the Neanderthals—were in their own way a highly evolved culture, people who were closely observing the heavens as long ago as 75,000 years ago and were capable of calculating cosmic events such as the long-term cycles of the moon and the periodicity of the planets. It also seems that not all of them were merely fighters and breeders, for they engaged in huge communal mining enterprises. Gooch cites as evidence for this large-scale red-ochre mines and quarries recently discovered in southern Africa. Some of the mines have true mining tunnels, and these immense hives of highly organized human activity have been accurately dated, with the earliest, around one hundred thousand years old, corresponding exactly with the long-term genealogist's latest possible date for the appearance of African Eve. It should also be noted that red ochre had no material value and so would have been used for ritual purposes only, indicating that the Neanderthal's world was not solely an arena of survival, but also had a strong spiritual side to it. This is further confirmed by the archaeological evidence mentioned previously, namely the stone altar discovered in a cave at Drachenloch in the Swiss Alps, in which had been enshrined seven bear skulls. Obviously seven is a crucial number here, not only because of its obvious association with the natural rhythms and harmonies of nature, but also because of its association with a celestial counterpart, the seven stars of the Great Bear. Even as early as 75,000 years ago, it seems, the Neanderthal consciousness was already nurturing the rudimentary elements of the theory of transcendental evolution, possibly little more than twenty or so thousand years after the appearance of African Eve. By about 35,000 BCE, Eve's descendants had migrated across the habitable regions of the Earth and all but wiped out the Neanderthal. Today only a few bones and tools are left to testify to their existence. Indeed, according to the American geneticist Mark Stoneking of the University of Berkeley, who has done extensive research on mitochondrial DNA, there aren't even any original Neanderthal genes left in mankind's gene pool. The evidence suggests that the modern human—Cro-magnon, *Homo sapiens sapiens*—was an extremely vigilant and ruthless exterminator. As Stoneking says, there are no non-African mitochondria in the genetic makeup of any individual living: "It looks like there wasn't any mating going on between the resident females and the migrating males—at least none that produced a lasting genetic legacy."[11]

However, as Gooch's research has shown, they certainly left a lasting symbolic, or metaphysical, legacy. We see this, in part, in the astronomical observations of both the Neanderthals and our forebears, the exterminators.

For example, the constellation of the Great Bear was called by the ancient Egyptians the Mother of Time and was later regarded in India as the heavenly home of the *septarishi,* an embodiment of the seven properties (*rishi*) of creation. We thus have a definite link between the mythological and astrological beliefs of the "primitive" hominid existing in the depths of the last ice age and those of the priest-astronomers of Old Kingdom Egypt and Vedic India. One would expect such links to exist, of course, because that's exactly how evolution works. Successful ideas are like successful genes and are passed on in exactly the same way, with or without the cognizance and cooperation of the host "organism." The symbolism of the number seven is one such "gene," and it has been with us since the dawn of civilization. Gooch not only identifies this number in the findings at Drachenloch, but also in another symbolic configuration known as the Cretan Labyrinth, which took the form of a sevenfold spiral leading to a central point. This design appears in both the pre-

Columbian Americas and in Minoan Crete—hence its name—which suggests that it originated from a common, extremely ancient source dating back, Gooch suggests, as much as 20,000 years. A simpler, less standardized design has been found on a Palaeolithic mammoth bone from Siberia, which again pushes its origins back at least as far as the time of the bear cult of the Swiss Alps, circa 75,000 years ago. This same "ritual labyrinth" is also found in the symbolic designs of Cornish, Rajastani, Hopi, Finnish, Welsh, and Etruscan art, all of which feature spirals with seven turns.

In Greek mythology, Theseus was told by Ariadne that the maze, or labyrinth, consisted of one left-handed, seven-coiled path spiraling in, and one right-handed path spiraling out—a kind of double-helix configuration through which the initiate "danced" his or her way to freedom—or to enlightenment.[12]

Now let's return for a moment to the conventional view of the hominid's recent development, that the archetypal caveman was superseded by the slightly smarter but equally cutthroat Cro-Magnon type around 35,000 years BCE. Some 25,000 years later, after enduring possibly the most treacherous and uninhabitable climatic conditions ever faced by mankind (the last ice-age meltdown), these people foregathered in and around the more temperate regions of the world.

Now contrast this hungry, flint-wielding creature with the kind of human being that lived in the Fertile Crescent about five thousand years ago. From Egypt to Mesopotamia there occurred, quite suddenly, an unprecedented bout of civilized activity. Great cities and highly refined cultures, the likes of which had supposedly never been seen before on Earth, grew and blossomed in the space of a few centuries. Suddenly man had learned to build on an unbelievable scale—not just any-old-how, but with an expertise and precision that even today's architects and engineers, using the same kinds of tools as the ancient builders are supposed to have used, would find very difficult to match, let alone surpass.

As is usual when trying to assess the incredible achievements of these "children of the hunter-gatherers," we shall take the classic example of the Great Pyramid, undoubtedly the largest and most complex building ever to have been constructed in stone. Each of its two and a half million or so blocks of limestone averages 2.5 tons in weight, and some of the larger granite blocks incorporated hundreds of feet up in its interior structure weigh as much as 70 tons. On top of all that, this monument, towering almost 150 meters above the Giza plateau, was originally sheathed in a two-and-a-half-meter-thick, exterior limestone casing of blocks weighing around fifteen tons apiece. These blocks were irregular in shape underneath and had to be made to fit the cruder contours of the core masonry, but, quite remarkably, their exposed surfaces were perfectly flat and polished so that they shone like glass. Further, each of these blocks, equivalent in weight to about thirty family-sized cars, was set with cemented joints only a fraction of a millimeter wide. Archaeologists and astro-archaeologists alike tell us that this remarkable building was erected around 2500 BCE, and all of the available evidence supports this dating.

So what was our primitive forebear doing for a living in the Fertile Crescent a millennium before this era? According to the current historical scenario, the natives had barely learned to tame grass, let alone their restless minds; they were merely surviving. But then, quite suddenly by evolutionary standards, the natives not only learned to build on an unprecedented scale, to write and to administer vast social enterprises, they also conjured up the remarkable Hermetic Code, pulling it out of nowhere, like a rabbit out of a hat, and then mankind suddenly came of age.

Now this, I would suggest, is a massive developmental leap. According to Darwinian theory, macromutations seldom occur in nature, and if they do they usually result in some kind of chronic deformity of the organism, which generally results in extinction. Yet here we have an example in nature of a genuine macromutation—in this case in the mind of man—despite the fact that it seems to break the fundamental Darwinian rule of natural, selective evolution.

So, what exactly happened? Is the Hermetic Code really the chance product of a random macromutation in the brain of a fortunate member of the species *Homo sapiens sapiens,* or was it purposely introduced by some kind of external force?

Possibly we shall never know. Some of the authors mentioned earlier have suggested that there existed, toward the end of the last ice age, around 15,000–11,000 BCE, a high civilization that was almost completely destroyed in a massive global catastrophe. Dozens of myths and legends from all over the world describe such an event, which apparently culminated in the Great Flood. An unprecedented rise in sea levels would have been a natural result of emergence from the last ice age, when there was a rapid deglaciation of vast regions of the Earth's surface.

The legends from America and Egypt all say that there was only a handful of survivors of this great cataclysm, seven or eight in number. In the Judaeo-Christian tradition these survivors are known as the Noahs, a seafaring people with extreme foresight who knew how to build and sail ships across oceans. They had also been initiated into the secrets of the Hermetic Code, as the biblical records clearly show.[13]

Elsewhere Noah and his companions (or similar survivors) were known by various other names: Osiris or Thoth in Egypt, Viracocha in Peru and Bolivia, Quetzalcoatl in Mexico, Yu the Great in China, Manu in Vedic India, Deukalion in Greece, Utnapishtin in Babylon— the list goes on, through more than seventy flood legends from cultures worldwide.

If these myths are in fact describing an actual event, then it is entirely possible that some of the survivors of an earlier civilization passed on the main tenets of their knowledge to the early settlers of the Fertile Crescent. So when ancient scriptures speak of "divine intervention" on the part of the "gods" from heaven above or whatever, possibly they are merely referring to the dissemination of a superior knowledge to a less advanced race, a perfectly logical transference of consciousness from spheres "above" to spheres "below."

Of course, the Darwinist theory of gradual change through chance mutations actually lends support to this idea that civilization is much older than historians would have us believe. Even so, the question of whether or not the concept of the Hermetic Code, however old it may be, originally evolved gradually or appeared quite suddenly in the fertile mind of a single inspirational genius remains unanswered. Perhaps the experiences of some modern "geniuses" can provide us with a clue here, for it is a well-known and accepted fact that scientists themselves very often experience moments of inspirational perception, intuitive insights that transcend logic.

A typical example is the strange experience of the German chemist August Kekulé, who, after spending several hours laboriously working through a mundane textbook, fell into a dispirited half-sleep, in which he saw long rows of atoms dancing before him, wriggling like snakes. When one of the snakes suddenly seized its own tail with his mouth, Kekulé awoke to the realization that he had just "seen" what he had been long seeking—the precise chemical structure of the benzene ring.

Similarly, Henri Poincaré, the French mathematician, said that the solution to a particularly difficult non-Euclidean geometry problem he had been grappling with came to him quite suddenly, at a time when he was idly thinking about something far removed from mathematics. Another great pioneering scientist, the astronomer Johannes Kepler, said that the discovery of his famous third law came to him as "a glimpse of light"; and Einstein, whose own "glimpses" into the mysteries of space and time are the stuff of modern legend, had this to say of scientific investigation: "There is no logical way to the discovery of these elemental laws. There is only the way of intuition."[14]

Perhaps, then, it was something like this flash of intuition (a macromutation of the mind) that was responsible for the conception of the Hermetic Code by Thoth/Osiris/Viracocha or whoever. Or it may be that the idea evolved gradually from the earlier instinctive impulses of the Neanderthals. We shall possibly never know exactly how this change came about, but in any event we can see that successful macromutations do, in fact, occur frequently in nature. A flash of intuition is precisely that.

As I said, Darwinists exclude macromutations from the evolutionary process on the grounds that gross physical changes are invariably detrimental to living organisms. And yet, regarding the origin of the very first intelligent entity to appear on this planet, the DNA–RNA complex, evolutionists' arguments for a gradual appearance are not entirely convincing.

Scientists now believe that the first biomolecular self-replicators were free-falling, bacterial RNA strands. Exactly what form their self-replicating predecessors took no one knows. Dawkins cites a proposal made by the biologist A. G. Cairns-Smith that the precursors to organic self-replicators might have been something like inorganic crystals, growing, say, in different sorts of clays, constantly transported every-which-way by ever-changing waterflows.

In fact, crystals do, in a sense, "grow," one into another, the first array of geometrically aligned atoms and molecules acting as a template for the next. As they grow, crystals also produce, on occasions, flaws (mutations) in their molecular structure, which are then "copied" by the subsequent developing layers. Crystals also possess right- and left-handed properties, that is, two varieties—two or more kinds being the necessary prerequisite for the phenomenon of heredity, where "like begets like." However, as Dawkins himself points out, crystal molecules only act as templates for the formation of molecules in their mirror image. So, in this particular instance, like does not beget like. Chemists have been trying for many years to "trick" inorganic molecules into breeding other molecules of the same handedness, but the natural forces at work in the inorganic molecular world are seemingly indifferent to such deception. If you start cultivation with a left-hander crystal, you end up with an equal number of left- and right-handed molecules. Thus, says Dawkins, "although the function of an earlier, non-organic self-replicator didn't involve 'handedness,' a version of this trick was pulled off naturally and spontaneously four thousand million years ago."[15]

It seems to me that this statement is somewhat lacking in scientific clarity. In fact, the suggestion that some kind of evolutionary "trick" was spontaneously "pulled off" all those years ago has a distinct air of the magician about it, a familiar, sleight-of-hand, "Hey, presto" quality, which suggests to me that its author is really a Creationist at heart, one who does believe in some form of "immaculate conception" taking place here on Earth way back at the dawn of geological time.

The analogy Dawkins uses in his book *The Blind Watchmaker* is that inorganic crystal growth, producing mutational flaws over billions of years, happened, quite accidentally, to act as a kind of crude "scaffolding" for the building of a sophisticated biochemical "arch." Once the final "center-stone" of the arch fell into place (one of the four bases, perhaps?), then the previously formed crystal "scaffolding," greatly superseded by its biomolecular successor, involved, or collapsed, into extinction. The now animated "arch," the biomolecular descendent of this extremely primitive inorganic ancestor, eventually evolved blindly into entities like Jesus or Einstein, Hermes, Dawkins, and ourselves.

According to Dawkins, "the digital revolution at the very core of life has dealt the final, killing blow to "vitalism"—vitalism being the apparently mistaken notion that living material is deeply distinct from nonliving material. This is certainly true in respect of the individual electrons and atoms of which living matter is composed, but when considering the overall symmetrical structure of entire biomolecules, and the harmonious distribution of the electrons and atoms within these beautiful, dynamic, musically structured life forms, then the fundamental difference between, for example, a molecule of the protein hemoglobin and a molecule of water is surely glaringly obvious. Ergo, vitalism—my kind at least—is very much alive and kicking.

We thus have two extremely advantageous macromutations in the otherwise uneventful story of the evolution of life on Earth: the quantum transition from salt crystals or whatever, via something or other, to living, writhing, hermetically composed bacteria; and the incredibly rapid metamorphosis, over a period of time which by Darwinian standards is infinitesimally small, of African Eve into Marie Curie or Rosalind Franklin.

But are successful macromutations really so very rare? Have they only ever happened at the beginning and at the end of this current evolutionary episode on planet Earth? Perhaps not. For example, the transition from the single chemical base to the amino acid is a huge developmental leap. Again, the transition from amino acid to protein molecule is also a massive step. And what about the transition from bacterial RNA to the first self-contained cell, or the first cell to a multicellular structure and so on, to the fish, the reptile, and the bird, the mammal, the hominid, and ultimately the civilization-builder?

Evolutionists will argue that these marked changes in development only look like macromutations, that they are in reality composed of untold billions of small, gradual, mutational steps. This may be so, but at some point in each of these evolutionary lines there must have come a point during the transition from one stage of development to another when a clear distinction between the two was finally cast. A bird is only a bird when it is a bird, not before. And we know that, at some point, birds definitely did come into being. When this happened, when the first feathered creature finally took flight, a greater macromutation occurred.

Perhaps the evolution of the Hermetic Code proceeded along similar lines, where a rudimentary instinctive awareness of the natural rhythms and harmonies of nature took root in the primitive consciousness and then gradually began to develop into a coherent belief system. One can envisage a scenario whereby this process of recognition might have continued to evolve to the point where all the fundamental components of the Hermetic Code were instinctively incorporated into ritual practices, possibly without any conscious intervention on the part of any single individual. At a certain stage, however, someone, or a group of people, must have realized that the separate components of this instinctively adopted number symbolism could be incorporated into an overall cohesive theory of evolution. Even if we don't know exactly how and when, we know this happened, because the Hermetic Code "happened." Gradualists, if they were to accept that the Hermetic Code is everything I say it is, or even that it exists, would probably argue for a slow dawning of this cosmic awareness, a step-by-step method of advancement.

But, as we have seen, there is plenty of room for any number of successful macromutations in the long, fragmented story of our evolution, and one cannot doubt that the "eureka" moment has been experienced untold times by millions of human minds for tens of thousands of years. And this process, as we have noted, eventually culminated in the emergence of the Hermetic Code.

If, as I believe, the Hermetic Code was a genuine macromutation, introduced by an extremely powerful external force into the receptive consciousness of the Fertile Crescent, then perhaps the first, hermetically composed biomolecular self-replicators, which evolved along exactly the same hermetic principles, also received, right at the very beginning of the evolutionary chain, an external "leg-up," to get them started.

What I am suggesting is that there may be very real, unseen forces in this universe that dictate that the evolution of DNA-based, or musically structured, life forms must inevitably, as and when conditions permit, occur everywhere in what is, after all, a musically structured arena. This is to say that the evolution of life on Earth, or anywhere else, far from being a chance, random event, is in fact an irresistible, natural process, dictated from start to finish by the natural forces of the universe itself. And these are forces that, as we have seen, are described in meticulous detail by the two fundamental laws of nature embodied in the Hermetic Code.

It is now generally believed that life, in some form or another, very probably exists elsewhere in the universe, though to what extent scientists can only speculate. The astronomer Frank Drake, in his book *Intelligent Life in Space,* formulated an equation designed to give some idea of the likelihood of life existing in other star-systems in our galaxy. The equation contains seven approximate factors, including the rate at which new stars form in our galaxy each year, the proportion of planetary systems that might harbor planets with a suitable physical environment, the smaller fraction of such planets on which life might actually get started, the number of years life is expected to survive on each planet—and so on. To get to the point, his answer, for our galaxy, is five-to-one against just one other planet in the Milky Way harboring any form of life whatsoever.

In the next chapter we shall be exploring this question of "alien" life in more detail. Being neither an astronomer nor mathematician, I will not be using long equations or intergalactic telephone numbers to explain why I believe that Drake has got his sums drastically wrong. In fact, all the reader needs to follow the line of thought to my own conclusions is some ordinary common sense, an essential pinch of intuition and, of course, a basic understanding of the Hermetic Code, the "theory of everything."

7

Extraterrestrial DNA

In this chapter we shall be considering some of the further implications of the theory of transcendental evolution. As we have seen, the neo-Darwinian theory of evolution is unfinished: it explains only the evolutionary development of organic bodies in the local biosphere of planet Earth. The theory of transcendental evolution, however, presents the whole picture, and it tells us that the process is continuing at an even higher level, beyond the confines of the physical brain of man, into scales of existence that ultimately encompass the whole universe. Before I try to explain how such a mechanism might work, however, we need first to get back to basics, to the fundamental components of the material world.

In previous chapters, we noted that the ancient Greeks had some rather unusual ideas concerning the nature of matter. They believed that all material things are "psychic"—alive—and that they are influenced in some fundamental way by music. Today we find that modern scientists have proved them right on both counts. They have discovered mindlike qualities in the electron and "organic" traits in plasmas, and they have identified eightfold musical symmetries in the two major physical scales of the microworld: the atomic and the chromodynamic. Then, of course, we have light, the eightfold symmetry of the white ray, with its curious twin photons that can simultaneously "feel" what the other is feeling, even if they are light years apart.

Further, we have seen that similar hermetic symmetries are also evident in the biomolecular world, with its sixty-four codon combinations and twenty-two evolutionary amino-acid signals. This means, therefore, that the whole of the microworld, from wave/particles to biomolecules, conforms very closely to the Greek view, which is that the entire universe is built, scale superimposed upon scale, of crystallized, ever-vibrant music.

We thus have three fundamental harmonies in evidence in the microworld: the chromodynamic, the atomic, and the (bio)molecular. Underlying all of these scales, of course, is the all-pervading harmony demonstrated by the "twin photon" phenomenon, the "actions at a distance" called by Roger Penrose "nonlocal quantum correlations." The whole universe is perpetually in motion and all wave/particles are continuously interacting and separating, which means that the nonlocal aspect of quantum systems is a general characteristic of nature. Clearly this represents, in the physical world, a harmony of the highest possible order. It is one thing to say that the universe is a harmonious entity because it is constructed entirely upon the eightfold chromodynamic and atomic matrices, but nonlocality suggests that there exists a far deeper interconnecting harmony underlying all physical phenomena, where everything is resonating at the very same subquantum frequency, everything is "in tune" with every other thing.

We now come to another very ancient idea, which again seems to have first surfaced in the time of the early Greeks: the notion that the whole universe is itself a living, sentient being.

The Greeks had a name for this creature, this universe. They called it the Zoon (pronounced "zoh-on"), the modern dictionary definition of which is "morphological individual, the total product of a fertilized ovum." Of course, the originators of this particular concept might not have defined it in such a precise way, but it seems obvious that they believed that the universe was an animal, living, hence the notion of zoology, the study of the living, of which Richard Dawkins himself is a professor. So the Greeks, I believe, regarded the cosmos as having somehow been conceived and then born, that it subsequently grew and is still growing, and that all systems within it, from the Earth and the planets to the sun and the stars beyond, are vital components in the living body of the whole.

As we know, the Greeks also believed, like the Egyptians before them, that the universe exists and operates according to musical principles, that is according to the fundamental laws described by the Hermetic Code. And the Hermetic Code, as we have seen, is identical in structure to the genetic code. This means that the universe, according to these traditions, is, in effect, an immense, multidimensional complex of evolving genes, that is, it is a biological entity. Let us note here that there is no ambiguity whatsoever in this ancient worldview. These people stated, in very clear and precise scientific (that is, musical) terms, that the Zoon is a zoon, so we may safely assume that they meant what they said.

Now this may seem like a tall order, asking us to believe that we exist in the living body of some mighty beast, some "god" of potentially infinite size. But then, not too long ago ideas about the hermetic symmetries involved in the creation of life, of "psychic" matter and of "universal music" would almost certainly have been regarded by scientists as being simply the fanciful products of primitives' dreams. And yet, ultimately, time has proved the ancients to be fundamentally correct in these particular aspects of their worldview; music is indeed everywhere, both inside and outside us, or "above" and "below," and the basic components of matter continually resonate and even communicate, being by no means truly inert or "lifeless."

So now these ancient thinkers once again reach forward through thousands of years of time to confront us with another strange idea, one that has never, as far as I am aware, been seriously considered either by alternative theorists or orthodox historians. This is the proposition that the whole universe is a living organism, quite literally, a biological entity.

I must admit that when I was initially confronted with this rather strange idea of a living universe I wasn't quite sure what to make of it. It was difficult to see how the greater components of the cosmos—such as all-consuming black holes, exploding supernovae, collapsing clouds of interstellar dust and all the other cataclysmic events taking place out there—could ever be construed as life. On the face of things, the proposition seemed absurd, or so I at first thought. But at the back of my mind I harbored a sneaking suspicion that these ancient people, who were so knowledgeable in other respects concerning the nature of reality, were not just simply fantasizing about their "living god," but had some basis for their belief.

So I began to investigate, to search for evidence of this. It seemed at first an impossible task. After all, if the universe is a sentient being, where is its "head," its "heart," and all the other organic components necessary for a body to exist? Answer: nowhere to be seen. Nevertheless I kept looking, scouring books on cosmology, astronomy, and astrophysics and the like. I learned quite a lot about the cosmos that was to stand me in good stead for what was to follow, but there were no obvious clues in the writings of modern scientists as to the possible nature of the creature I was searching for.

But then, after long deliberating and vacantly scratching my head, I suddenly experienced one of those familiar "eureka" moments when, out of the blue, a new perspective dawned in my mind. Not surprisingly, perhaps, the answer—or at least, a major clue on its trail—came not from the modern scientist but from Hermes himself.

The clue lies in the now familiar saying of Thoth/Hermes, "As above, so below." What this undoubtedly means is that the world above—the greater cosmos—is fundamentally the same as the world below, the world of man. Obviously the scales are vastly different, but, according to Hermes, their inherent structure is based on the same hermetic blueprint. And remember, the Hermetic Code and the genetic code are also identical in every way, so we can see quite clearly that, in respect to the world of man and the world of the cell, the hermetic dictum just quoted is directly and exactly applicable. The genetic code operates in the microcosm, the world of cells, and the Hermetic Code operates in the mesocosm, the world of man. And each of them, of course, shares a common purpose, which is to facilitate the processes of creation, of evolution.

Now DNA is quite clearly the prime mover in the biomolecular world. It is DNA that employs the genetic code to manufacture amino acids, which other organic components then assemble into proteins. The purpose of proteins in the biomolecular world is clear; they engineer all the complex chemical processes that build a living body. So what happens when the human mind employs the Hermetic Code? Does it, as I have already suggested, produce the metaphysical equivalent of amino acids? Possibly. At least it produces ideas, thoughts, and concepts, which are born of our conscious ability to emote, sense, and perceive.

So on to the obvious question: are these concepts integral parts of a much bigger evolutionary process that takes place somewhere "out there"? That is, if the genetic code describes an organic process, is not the Hermetic Code similarly describing an organic process, but one that operates on a much greater and more rarefied scale? If this is the case—and I am, of course, proposing that it could be—then there would probably be in existence other, greater, organic components out there in the macrocosm that could somehow assemble these metaphysical "amino acids" (thoughts, concepts, and so forth) into the conceptual equivalent of protein chains. As I said moments ago, the purpose of proteins in the biomolecular world is to engineer all the complex chemical processes that are necessary to build a living body.

As the reader will by now have realized, what is being implied here is that the human mind is a form of "double helix," a chromosome in the nucleus of a living cell in the body of a much greater being. And then, further, possibly the "mind" of this greater being is also but a single-cell nucleus in a being on an even greater scale . . . and so on, but not, as we shall see, ad infinitum.

So let us now follow this lead as we were probably intended to and turn our attention skyward, toward the heavens. If Thoth/Hermes is right, one would expect there to be signs of this "extraterrestrial" life up there in the macrocosm—massive, cosmic, organic structures, not unlike the structure of our chromosomes. And what do we find up there in the greater cosmos? Significantly, there are spirals and helices, literally everywhere, in all potential solar systems, in all galaxies. In the case of our own solar system, we see the planets encircling the sun, but as the entire system is perpetually moving at great velocity through space, the path traced by each planet is, in fact, a spiral. Similarly, most galaxies are spiral galaxies, which is meaningful in itself, but even so-called elliptical and irregular galaxies all revolve around a galactic center—maybe a black hole—and all of them are hurtling across the universe at tremendous speeds, so the trajectory traced through space-time by every star is a true spiral, a helix. Furthermore, we ourselves, as we sit, walk, or even sleep on the surface of the Earth, are actually tracing spirals through space: the whole planet spins like a top as it moves ever onward.

It is true, of course, that DNA is a double-helix structure, but in the biomolecular world there are many single-helix structures—viruses, bacteria, and so forth—composed of single strands of DNA's close cousin, RNA. Moreover, when we think we see a single spiral out there in space, we are not necessarily taking in the whole picture. That is, there may be "invisible" helices that we also need to identify. In a typical galaxy, for example, scientists have discovered that the visible spiral arms, composed of stars—of light—are enveloped by an accompanying magnetic field that actually spirals around each of the arms. Again, in the solar system there are several planets, each one spiraling along its respective trajectory, but if you consider the combined paths of any two of them in relation to one another, the result, quite clearly, would be a double-helix configuration. The point is, it is the helix itself, whether single, double, or even multiple, which appears to be the basic design for all evolutionary phenomena, above and below.

Is it not highly significant, therefore, that the Egyptians and the Greeks, in whose belief systems the firmament above was so important, should have a principal god of wisdom whose symbol was a magic wand known as the caduceus, featuring a double helix in the form of two entwined serpents surmounted with wings? Clearly this symbol, like the Great Pyramid, which is aligned so precisely with key stars in the Duat (sky), is inducing us to look heavenward. The same can be said of the serpent/thunderbolt symbols of Viracocha and the plumed serpent of Quetzalcoatl, of which the most impressive depiction of all is the effect at the spring and autumn equinoxes of the sun's light undulating like a serpent up the northern staircase of the Temple of Kukulkan at Chichén Itzá, Mexico.

After I had identified these "serpents in the sky" it seemed to me that the most logical thing to do would be to be to try to follow their trail and see where it might lead. This was, after all, the route taken by Osiris and Thoth, and all of the other principal "civilizers" of the ancient world.

The fact that Sirius and Zeta Orionis, the two most important stars in Egyptian cosmology, are targeted by the southern shafts of the two principal chambers of the Great Pyramid suggests that the pyramid itself provides a vital link between worlds above and below. The name of the Great Pyramid—"The Lights"—suggests further that light is the cosmic intermediary, the interface between man and god, the earth and the stars. And light itself, as we have noted, is a musical phenomenon: the most resonant octave in existence. Therefore music is the key. But then music is a hermetic phenomenon, and the Hermetic Code describes exactly how the genetic code operates, and how organisms grow.

What I believe is implied here is that we, in our quest to discover the secrets of the universe, have literally to evolve our way to the stars, to grow so as to be able to touch the firmament merely by holding out our hands. The kind of "growth" we are talking about here, of course, is the evolution of the mind, an "alchemical" process of development that leads ultimately to a condition of optimum psychological harmony and an awareness of the nonlocal realm of heaven, the timeless plane of light. Here, Osiris/Orion is everyone's immediate neighbor.

So let's now try to envisage how this spiritual growth might develop. By this I mean the process by which consciousness can eventually tune in to the nonlocal dimension, or to the metaphysical frequencies characteristic of light quanta.

We have seen how, in the microcosm, the most powerful transmitter of intelligent data is the DNA double helix, which codes for complex biochemical processes with such precision that it can create the brain of a human being. In the microworld, this evolutionary music played by DNA represents a harmony of the highest possible order.

The human brain, the ultimate product of DNA's evolutionary development, is more than just a biological organ like a heart or a liver. It possesses self-awareness and can perform a whole range of extra-biological or metaphysical functions—intellectual, speculative, intuitive, or whatever. This is to say, the human brain, like the double helix from which it originates, can generate transcendental influences. Through literature, artifacts, buildings, and so on, it can transmit "biometaphysical" signals to spheres far removed from the physical body in which it exists. Thus the designers of the Great Pyramid, for example, or the I Ching, or the authors of the Christian Scriptures, the Koran, or the Upanishads, living in the remote past in distant parts of the world, are still speaking to you now through these and many other contemporary commentaries. These works are, in effect, transcendental phenomena, metaphysical genes, coded to synthesize—in the mind of the human being—higher, more complex, modes of cognition.

As we can see, both the DNA double helix and the human brain function in strikingly similar ways. In the concluding chapter of *The Infinite Harmony* I suggested that the conscious/subconscious aspects of the human brain, with its right and left hemispheres, could be regarded as the metaphysical equivalent of the acid/alkaline aspects of the DNA strand, with its "right" and "left" nucleotide chains. Both work with the same hermetic/genetic blueprint, with its four "bases," its sixty-four possibilities and its twenty-two transcendental or evolutionary "signals." Thus the music being played in the microcosmic processes of evolution is being echoed, note for note, in the mesocosmic scale above, in which the human mind is the principal player. So the ancient dictum "As above, so below" means, quite literally, that the only difference between the DNA double helix below—the chromosome—and the fully functioning human brain above is simply one of scale. Both work with exactly the same numbers and combinations of forces and components, but the components themselves are graded accordingly, the bases used by DNA presumably being of a less rarefied form of resonance than the "bases" from which higher frequencies of consciousness are developed.

So what exactly are these "bases" from which consciousness is created? In the Hermetic Code they are symbolized by the four base notes—the four Dos—of the triple octave, which correspond to the four base pebbles of the Pythagorean Tetrad discussed in the last chapter. Now these bases, like the free-floating chemical bases in the living cell, would theoretically be everywhere, all around us, waiting to be scooped up by the double helix of the mind, combined into more resonant units, and finally passed on for future synthesis at a higher level of development. These bases, I would suggest, come to us in the form of our impressions, namely our sensations, emotions, and perceptions, the trinity, or triple octave, within us all. There is a fourth, of course, a crucial, transcendental base, the product of the harmonious interaction of the first three, which, if fully developed, manifests in the form of our concepts, our conceptions.

We have here, I think, the origin of the much misunderstood Christian notion of the "immaculate conception." Jesus's teachings, for example, in being psychologically harmonious, were immaculately conceived, born of a conscious, highly developed mind. We know they are immaculate because, just like the equally resilient concepts of, say, Pythagoras, the Buddha, or Muhammad, they have ultimately harmonized with the evolving psyches of billions of individuals.

You could, of course, argue, as Dawkins might, that these world-wide religious and philosophical movements have evolved purely by chance and that they have played on the inherent weaknesses of desperate human beings wishing to escape the tedium and pain of Darwinian existence by clinging on to vague, unfounded promises of an afterlife in Paradise. Doubtless the element of escapism is a contributory factor in the development of many self-help cultures, past and present, but the fact that the world's major religious movements are all based on the principles of "esoteric music" embodied in the Hermetic Code strongly suggests that accident really has very little to do with it. Just think for a moment: What are the chances of your ideas, or those of Darwin, Einstein, or Dawkins (or your own modern hero or heroine), entering and attuning with the minds and hearts of whole races of people?

Significantly, in the microworld of the cell, "immaculate conceptions," or macromutations, are the very life-blood of creation. All evolutionary advancements, all transcendental developments, are "immaculately" created: amino acids, protein molecules, eyes, legs, wings, brains. So of course Jesus was immaculately conceived. But then so was African Eve, *Tyrannosaurus rex,* and, paradoxically, the AIDS virus. Life is like that.

So if the human brain is a form of chromosome, a metaphysical double helix, then presumably it is also an integral component—a "cell nucleus"—in the greater body of an infinitely more complex, macrocosmic "organism," a creature that, one assumes, is formed from the collective evolutionary consciousness of the entire human race.

As a matter of fact, a scenario not dissimilar to this was put forward in the early 1950s by the American writer Rodney Collin in his book *The Theory of Celestial Influence*. I have quoted at length from this unique work elsewhere. Collin was a close associate of Ouspensky, and he spent several years compiling this scientific interpretation of Gurdjieff's original ideas. The book has not yet received the worldwide acceptance long overdue to it, but I believe the time will come when the writings of all three of these highly innovative thinkers will be recognized as great achievements.

In the final chapter of his book, entitled "Man in Eternity," Collin tries to imagine the form and structure of the greater "body" of the human race. Everyone, he says, creates an invisible thread of converging energies as he or she lives along their own particular line of time, each thread being unique to a human being. If one imagines, across the centuries, billions upon billions of these threads, crossing and interlacing with one another, varying in "color" and intensity according to the kind of life lived, then there emerges a figure so intricate that it is, in fact, a solid, the "solid of humanity":

> Of this solid we can even have a certain vague apprehension. It will be, as it were, a sort of solid tapestry, composed of billions of threads, which in spite of their inconceivably elaborate weaving, appear all to lie in the same direction which is eternity. We can even suppose each of these threads to have a different nature or color, according to the level of energy which dominates its totality of lives. And we might find that in large areas or periods of humanity, a certain nature or color dominates the whole design—the red of purely physical existence, the yellow of intellectual activity, or the green of moving skill and sensation. Remembering the existence of men with conscious souls, and with conscious spirits, we shall also suppose threads of different materiality which stand out from the fabric in a quite exceptional way, which impart life to the rest, and about which the whole design of the solid body is formed. For those threads are threads only in our metaphor. In fact they are alive and their total mass is alive. They are the cells and capillaries and nerves of a body, the Adam Kadmon of The Kabala, Mankind.[1]

Adam Kadmon has appeared elsewhere in ancient mythology. He is the titan Atlas of the Greeks, who, remember, believed that the human race could change the world if it could harmonize itself into an overall coherent state of homonoia. And if the earlier-discussed evidence for psychokinesis is accepted, together with the recognized power of "group consciousness," by which a fully grown individual can be lifted with very little physical effort, then we can say that the kind of "resonances" obtained in such an orchestrated process of thought are real, they must have some sort of substance. According to the theory of transcendental evolution, this substance, the living stuff of consciousness, is created in precisely the same way as are all other manifestations of life, that is hermetically, genetically. Thus evolutionary consciousness itself is quite literally composed of metaphysical signals, or "notes," which have been copied from "genes" housed in the electrochemical structure of other mesocosmic "helices," other human brains.

The characteristic spiral form of DNA has been photographed in its totality through a process known as X-ray diffraction. It is clearly a double helix, and all DNA molecules in every living plant or animal are structured in exactly the same way. This DNA does not suddenly appear fully formed: it develops in a linear fashion over a given period of time. As the parent DNA ladder "unzips" at one end prior to cell division, and free-floating bases link up with the open ends of the split rungs, two identical chromosomes are formed.[2]

It might be argued that the description "double helix" cannot reasonably be applied to the two hemispheres of the brain, which look more like three-dimensional segments of some weird exotic fruit than the structural features of a chromosome. But of course, like DNA, we must assume that this metaphysical "chromosome" does not appear ready-made, but requires a given period of time to develop its overall form, and that this unfolds in a linear fashion. Time is the line and the period in question is a lifetime. And what exactly happens during the lifetime of an individual brain? Remember that the two hemispheres have a physical existence on the surface of a planet that is spinning endlessly on its axis as it soars through space. Therefore, each hemisphere is in fact tracing a spiral, a helix, through space and time. Taken together the overall configuration is, of course, a double helix.

So the human brain is an evolving double helix developing in time, at one end of which lies conception, at the other, death. All that happens between, all our experiences in life, both conscious and subconscious, define the particular "color," or quality, of each evolving mesocosmic "chromosome." The most successful or the most resonant of the "genes" in these chromosomes, like, say, the "bright," enduring ideas and concepts of those such as the Buddha, Christ, or Muhammad, are those that are replicated most, in succeeding generations, by other metaphysical chromosomes, other human brains.

The DNA structure of living cells has two sides to it, two chains made up of sugar, phosphate, and nitrogenous bases, each side being a mirror image of the other. The two chains are held together by the bases: adenine, thymine, cytosine, and guanine. Adenine always links up with thymine, and guanine always joins with cytosine. Similarly the human brain, as we have seen, also has two sides to it, the conscious processes of the left hemisphere and the subconscious processes of the right. As with DNA, these different aspects of the two hemispheres are inextricably linked to one another by "bases," by our experiences in life, born of our sensations, emotions, perceptions, and conceptions.

This, then, forms the core of the ancients' view of mankind's evolution. Consciousness is an organic phenomenon, it develops according to the dictates of the Hermetic Code, and its natural inclination is to grow upward and outward, toward the stellar scale of existence, the nonlocal plane of light. So as consciousness evolves, the whole of the human race, like the pyramid builders of the ancient world, will be working together in states of ever-increasing harmony to build, in a higher dimension, a vastly more complex, macrocosmic structure. Theoretically such an entity would either possess, or would be evolving, a "brain" of its own, an immense "solar" helix that, in terms of scale and complexity, would be as far removed from the individual mesocosmic double helix—the human brain—as the mesocosmic double helix is from microcosmic DNA. If this is true, then we should expect to find the main components of this helical structure in the cosmos above.

We already have a clue as to the manner in which this awesome, extraterrestrial life form might evolve. This arises from a suggestion I made previously, which is that the creative processes in the metaphysical world of the human mind involve some kind of interplay between light and consciousness. This is to say that these two complementary yet quite distinct phenomena are the main components of conception/creation, opposite sides, as it were, of the same metaphysical coin.

Excluding starlight, the source of all light in our solar system is the sun. And the source of all consciousness, as we know it, is the Earth. Thus we see that this macrocosmic structure, exactly like its micro- and mesocosmic counterparts, has two fundamental aspects to it.

In the case of the DNA molecule, its two main characteristics are its acidic and its alkaline properties. Its function is to take in data in the form of individual chemical bases, which it then transmutes "up" into the more resonant RNA components, finally transmitting them back out into the cytoplasm, the liquid membrane of the cell, for future synthesis. Similarly the human brain, with its conscious and subconscious characteristics, also takes in data—impressions, perceptions, and so forth—that it then transmutes into a kind of metaphysical "light," subsequently radiated out into the world in the form of ideas, concepts, and the like.

So, with regard to the double helix in the sky, we might say that the sun is its "acidic," or conscious aspect, and that life on planet Earth is the manifestation of its "alkaline" or subconscious aspect. Presumably the celestial double helix too would take in data of some kind, which it would then transmute up and radiate out, but exactly what form this might take is perhaps a question that only Adam Kadmon himself could answer. However, we can speculate. If there are macrocosmic "bases" up there, they must be pretty big, and the biggest components of the solar system are the other planets and asteroids, all of which radiate some kind of magnetic influence out into the greater system.

Significantly, the ancient Greek pantheon of the gods, Zeus and his celestial family, was closely associated with the sun and the planets, each of which was regarded as being imbued with life. They also referred frequently to the "music of the spheres," believing that the whole solar system is a hermetic creation. And hermetic, of course, is genetic, organic.

It so happens that this "music of the spheres" is not simply folk-lore, but fact. Rodney Collin noted that the major and minor con-junctions of the planets all "beat out" certain rhythms that can be numerically defined in a regular sequence of harmonic intervals developing in time. When he subsequently made a comparative table of these conjunctions, he discovered that the figures obtained, taken as vibrations, represent the relative values of the fundamental notes of the major scale.

As he says, the periodicity of a planet's magnetic influence follows the time necessary for it to return to the same relative position of clos-est proximity to Earth. Collin takes as the point of departure of each planetary cycle the moment when the sun, the Earth, and the given planet are in a straight line. The cycle of that planet is thus the time that elapses before such a conjunction occurs again; it is the "interval" between the recurring moment when the three magnetic forces of sun, Earth, and the given planet act together in the same way.

Mercury and Venus repeat their maximum magnetic effect every eight years, the asteroids every nine, Jupiter twelve, Mars fifteen, and Saturn thirty. When these various rhythms are superimposed one sees an interesting sequence of harmonic intervals, each stage of which is marked by the major conjunctions of one or more planets. Every twenty-four years Jupiter completes two full cycles, Venus and Mercury, three each. The next significant stage occurs every twenty-seven years, at which the asteroids complete three full cycles. Every thirty years Mars completes two cycles and Saturn one. Every thirty-two years, Venus and Mercury each complete four cycles. Every thirty-six years, Jupiter runs through three cycles and the asteroids complete four. The next stage is forty years, during which Venus and Mercury each run through five cycles. Every forty-five years Mars completes three cycles and the aster-oids five. Finally, every forty-eight years, Jupiter completes four cycles and Venus and Mercury each complete six full cycles. This sequence is constant, repeating itself endlessly.

We thus have a table of the numbers of years, beginning at 24, rising up through 27, 30, 36, 40, 45—and ending on 48, exactly double the value of the number we began with. This is significant, for not only does this series of numbers conform to the relative values of the notes of a major scale, exactly like an octave of music, it also expresses the ratio 1:2. Collin concludes: "And we are reminded of old stories that this same musical scale, ascribed by legend to the Pythagoreans, was invented by a special school of astronomers and physicists, to echo the music of the spheres."

When we tried to envisage the "double helix" formed from the two hemispheres of the human brain as they interact along the line of time, we noted that the true form of the whole evolving phenomenon spans the life of the individual, from conception to death. The "double helix" of the solar system exists in an even greater dimension, and so in terms of size and duration of existence must be as far removed from the mesocosmic scale of the human brain as the human brain is from the microcosmic scale of the individual cell's chromosomes. Consequently we must try to view the basic chromosomal structure of the "solar being" in relation to the cycles of the planets, and in particular of Earth. We might say, therefore, that this macrocosmic "chromosome" started to develop when *Homo sapiens sapiens* first began thinking in simple concepts and that it will continue until such time as consciousness on Earth ceases to exist. It is about one hundred thousand years or so since the Neanderthal or early Cro-Magnon first started mining for red ochre in southern Africa, not long after the beginning of the last ice age. It is believed that the ochre was for ritual rather than practical purposes, suggesting that the evolution of "consciousness" was well under way by then. So the solar being of mankind is at least a hundred millennia old.

Over this expanse of time, the helical patterns traced by the components of the solar system as it spins like a giant Catherine wheel through space form an immensely long and complex figure. Collin describes this four-dimensional structure in his own unique style:

> The planetary paths, drawn out into manifold spirals of various tensions and diameters, have now become a series of iridescent sheaths veiling the white-hot thread of the sun, each shimmering with its own characteristic color and sheen, the whole meshed throughout by a gossamer-fine web woven from the eccentric paths of innumerable asteroids and comets, glowing with some sense of living warmth and ringing with an incredibly subtle and harmonious music.[3]

Collin did not identify this structure as a "chromosome" as such, but the reference made to its living warmth was certainly intended to be taken literally. As he says a little further on, "the solar-system is, in some way incomprehensible to us, a living body."

The Greeks, as we have seen, shared much the same view, but they also believed that this immense, musically structured entity is a conscious being with a mind of its own. Presumably this would be a mind that, like ours, takes in external data—impressions of some kind—and out of them constructs the macrocosmic equivalent of concepts. Clearly the possible nature of such data is incomprehensible to us. If the body of the solar being is constructed from metaphysical "amino acids," that is from our concepts coupled with light, then the "mind" of the entity would inevitably function with much higher, more rarefied energies, perhaps operating at velocities far greater than the speed of light. According to Special Relativity, of course, nothing can travel faster than light. Physicists have in the past tried to overcome this limitation by dreaming up a new kind of particle, a "tachyon," a hypothetical entity, existing in a higher dimension, that travels backward in time at speeds greater than the velocity of light, but never below it. Theoretically the tachyon cannot exist, but even if it did, we might never know it, because any particles moving (backward in time) at such a phenomenal speed could never be detected by any known scientific means.

At any rate, if the solar "mind" is a reality, then there must be some form of energy sustaining it. And if light is the limiting factor in the mesocosm, then perhaps the limiting factor in the greater scale above is in some way directly related to it.

As it turns out, in the Hermetic Code we have a clue to the possible relationship between these two scales. The key number is 64, the number of amino-acid templates in a living cell, the number of hexagrams in the I Ching. Sixty-four is the square of the constant number 8—8 being the symbol of the octave, the fundamental component of *pi,* and the basic matrix of all creative processes.

Remember also that this number was closely associated with the Great Pyramid, which was itself closely associated not only with the phenomenon of light but also with the stars, with the stellar scale above. It seems to me very unlikely that the symbolism involved in the whole pyramid phenomenon should be accidental or arbitrary. The details are too precise for that. We have the Great Pyramid itself, "The Lights," designed and constructed by the followers of the god of wisdom, one in a pantheon of eight gods, whose symbol was two entwined plumed serpents, and whose "Magic Square" embodies all the numbers from 1 to 64. Surely we are being told here that the key to transcendental evolution is the square of the constant, that the sacred metamorphosis from man to god described in the pyramid ritual represents the squaring of one's possibilities. Possibly this is why the ancient names for the Great Pyramid are in the plural: Khuti, "The Lights," in Egyptian and Urim middin; "Lights-measures" in Chaldee and Hebrew.

So, while modern science denies the possibility of superluminal (faster-than-light) motion, the ancients' description of cosmic events strongly suggests that, in one way or another, it is an attainable reality. Of course the limitations imposed by Special Relativity refer to matter as we know it, the smallest components of which are subatomic or virtual wave/particles. However, the scale of materiality need not necessarily end with the wave/particle. Readers may recall that the physicist David Bohm's research led him to conclude that consciousness itself is a form of matter. But if consciousness is a form of materiality, it is clearly one of an entirely different order than the kind we can detect or measure, and so may not be bound by the normal laws of physics.

Now there are, in fact, phenomena in existence, such as correlated photons, for example, the nonlocal connectedness of which produces in the observer the illusion that something passing between them is "moving" at speeds far greater even than the square of the constant velocity. We shall return to this question of nonlocal communication later, as I believe it may provide a mechanism by which information can be transferred across the entire universe, from one macrocosmic "organism" to another, in less time than it takes to blink.

For the moment, however, we can tentatively hold on to this idea: the value of the square of the speed of light may be the diffusion speed of the interactive "resonances" of the solar mind.

Now we must move on—always upward, of course—and continue the incredible journey first made by the serpent gods of wisdom.

8

Interstellar Genes and the Galactic Double Helix

In the four-dimensional structure of the solar system's long body we have seen how the planetary trails, all encircling the white-hot thread of the sun, form immense helices in space. If we imagine each of these individual sheaths to be coupled in some way with the greater spiral motion of the sun, as there are nine planets, we can say that there are nine "solar" helices. One of these, formed by the combined motion of the Earth and the sun, is fundamentally different from all the others, in that it contains life, consciousness, you, and me. This particular double helix is the "brain" of the solar being, the mind of Adam Kadmon.

But, of course, if this solar being is indeed organic, a "chromosome," then, theoretically, like all helices it too would have the capacity, in a still higher dimension, to create, to build even greater, more complex forms of life. So, just as DNA forms the nucleus of a cell in a physical body, and the human brain forms the nucleus of a cell in the solar body of mankind, so too would the mind of this solar being form the nucleus of a cell in the greater, galactic body. This further quantum leap, from the scale of the solar system to the vaster galactic scale, means that the human brain, earlier defined as the mesocosmic double helix between DNA below and the solar configuration above, now becomes a microcosmic entity in a yet larger existence, in which the solar being represents the mesocosmic creative force, and the galactic body the macrocosmic. We shall take a closer look at these relative scales later in this chapter.

In the last chapter we noted that Rodney Collin had discovered that the orbital cycles in the planets of the solar system produce major and minor conjunctions in time, whose relative values correspond very closely to the harmonic proportions of the major scale. This legendary "music of the spheres" was frequently alluded to by the writers of ancient Greece: it was referred to by Plato as the "song of the sirens." The hermetic symmetry of these planetary motions, as I have suggested, is an indication that there are genetic, organic processes operating in the planetary sphere.

It so happens that, very recently, further evidence has come to light concerning the relationship between the masses of certain stars that seems to indicate that this planetary harmony may extend far beyond the solar system, out into the galaxy.

While I was working through the second draft of this book, which did not then include what follows, I received a phone call from Colin Wilson. He said he had been asked to review an updated edition of a book by Robert Temple called *The Sirius Mystery,* first published in 1976, in which there was some very interesting cosmological data that he felt would be of interest to me. Colin had already seen a hastily written first draft of this book and had been kind enough to offer some suggestions as to its presentation and format. So he knew exactly where I was coming from and promptly realized the relevance of Temple's conclusions to my work. He duly sent me a copy of *The Sirius Mystery,* which I had first read many years ago, but this new edition, as he had promised, proved to be very interesting indeed.

Temple is the man who introduced to the world the Dogon tribe of Africa, whose ancient and secret traditions contain very precise astrophysical data about Sirius and two other invisible stars in the Sirius system that have only been discovered in recent times. These two hidden companions of Sirius are known respectively as Sirius B, a white dwarf star first photographed in 1970 by Irving W. Lindenblad of the U.S. Naval Observatory in Washington, D.C.,[1] and Sirius C, a red dwarf star whose existence was only officially confirmed in 1995 by the French astronomers J. L. Duvent and Daniel Benest.[2] The Dogon, it appears, were well ahead of their time.

Temple believes, reasonably enough, that this knowledge came to them in the remote past, probably from Egypt. Then he advances the theory that the Egyptians and the Sumerians obtained this knowledge directly from highly advanced amphibious extraterrestrials from the Sirius star system. He cites as part of his evidence the prominence in certain myths of amphibious creatures, half-man, half-fish, who were said to have founded the first civilizations in the Fertile Crescent. The leading "fish deity" was known under various names on the eastern flank of the Crescent, although in Egypt there is no major god answering to the description given. In Babylon and Assyria this god was known as Oannes (possibly an early form of the name John); in Sumeria, Enki; and to the Dogon tribe in Africa, Nommo.

I have to say that Temple's idea of amphibious spacemen flying in from Sirius with their superior wisdom is not my favorite explanation for the birth of Earthling civilization. A more plausible theory is that the ancient civilizers appearing in all the major myths, said to have survived a Great Flood, landed on the shores of their new homeland in boats—hence the emphasis on the element of water. This proposition is further supported by the theory of transcendental evolution itself, as interpreted in many ancient legends, in which the basic element of water is primarily an evolutionary symbol, expressing the central importance of the passive, "watery" element in the process of creation (see chapter 9).

So, according to the theory of transcendental evolution, superior or "extraterrestrial" intelligence actually develops from below. And it grows, evolves, organically, ever upward, toward the stars. Temple's view therefore appears to be upside down. His arguments in support of his theory are extensive, suggesting technical maneuvers on the part of the "fish gods" that defy comparison with anything ever accomplished on Earth, including the construction of the Great Pyramid. These include the use of water-filled spaceships capable of interstellar flight, and also the construction of Phoebe, the smooth-surfaced, tenth moon of Saturn, which Temple believes may be an artificial, water-filled satellite constructed, or perhaps "inflated," by these fishlike creatures and used as a kind of staging post on their intermittent journeys to and from Earth.

Notwithstanding our obvious differences concerning the true nature of "alien" life, Temple has discovered some interesting new facts concerning the Sirius system and our own sun, which appear to link both star systems with the Giza plateau, in particular with the Great Pyramid and the second Pyramid of Khafre (Chephren).

As the Great Pyramid has an apparent Sirius connection (that is, the southern shaft emanating from the Queen's Chamber, which targeted Sirius as it culminated at the meridian at the time of construction of the Great Pyramid), Temple proposes that it might be a representation of the "invisible" star, Sirius B, and that the slightly smaller Pyramid of Khafre represents our own sun. This view might appear to fly in the face of the suggestion made by Robert Bauval that the three Giza Pyramids represent the three stars of Orion's Belt. The whole necropolis, however, as we have seen, is extraordinarily multifaceted, so it would hardly be surprising if we were to find yet more information relating to the Sirius system encoded within the design.

Temple begins by comparing the sides of the slightly larger base of the Great Pyramid (755.79 feet) with the sides of the base of the Pyramid of Khafre (707.75 feet), calculating that the sides of the Great Pyramid are 1.0678 times those of Khafre's. He then notes, using the newest available astrophysical data, that the mass of Sirius B is 1.053 times the mass of our sun. As he says:

The correspondence is thus accurate to 0.014. However, even this tiny discrepancy may be highly significant. For 0.0136 (which rounded off is 0.014) is the precise discrepancy between the mathematics of the octave and the mathematics of the fifth in harmonic theory, where 1.0136 is referred to as the Comma of Pythagoras, and was known to the ancient Greeks, who are said to have obtained knowledge of it from Egypt.[3]

As a matter of fact, I have already discussed the Pythagorean Comma in my earlier book, in which I proposed that it was intended to highlight the fundamental difference between ordinary, practical music and what I call "esoteric" music—ordinary music, I believe, having been considered by the Pythagoreans as being slightly "off-key" from the true harmonic constant from which life is created.

Temple expresses much the same idea in his revised version of *The Sirius Mystery,* in which he calls the discrepancy of 0.0136 (rounded off to 0.014) the Particle of Pythagoras: "Essentially, one could say that it expresses the minute discrepancy between the ideal and the real."[4]

Temple's "ideal" music in this context is what I would call "esoteric." "Real" music therefore is ordinary practical music. The harmonic deviation described by the comma is significant, raising the wider issue of how this discrepancy might have been rectified by the Pythagoreans: how they transformed ordinary music into what Temple calls the "ideal" kind. I have dealt with this in some detail in *The Infinite Harmony,* where I suggest that the marginal imperfection of ordinary music was connected with the "glitches" of the major scale. As I pointed out in the introduction of this book when introducing Gurdjieff's exposition of the law of octaves, these "glitches" are identified as the two points in the octave where the rate of increase in pitch frequency between one note and the next retards, that is where there are not full tones but only half-tones: between the notes mi–fa and ti–Do. This inherent deviation in the line of development of an octave in ordinary, practical music is the underlying pattern of development of all natural phenomena, and accounts for the vast multiplicity and variety of physical forms in the universe. Thus, while the music of our favorite composers and artists sounds perfect to our ears, the Pythagorean Comma indicates that it is never quite so.

"Ideal" music, however, the esoteric music of the Greeks, is organic music, the music of the Hermetic Code and the genetic code. This very special kind of music actually takes account of, and rectifies, the discrepancy highlighted by the Comma. Essentially, of course, this is the "music of the mind," the music from which life itself is created. It is Egyptian alchemy, which involved the application of the law of octaves as a mode of being, but with a very slight yet crucial additional input in each developing scale at precisely the two semitone points mentioned above. This means that a fully developed "psychological" or organic octave is composed not of seven stages, as in a normal scale, but of nine, because it includes within it the two extra impulses at the points of the missing semitones. If we remember that each of these nine stages, according to the second fundamental law of nature, is itself an octave, then quite clearly we have a genuinely perfect scale consisting of sixty-four "inner notes" (9 x 7 + 1, the 1 being the final Do).

Readers wishing to explore in greater depth the theory of the "missing semitones" may care to consult the relevant section of my previous book,[5] but for the present we must return to the main cosmological theme of this chapter.

Temple goes on to reveal that the precise value of 1.053, which we have noted has only very recently been identified as the exact ratio of the masses of Sirius B and our own sun, was very accurately expressed by the astronomer/mathematician Macrobius in the fifth century CE in the form of the "sacred" fraction, 256/243. Macrobius claimed that this fraction, which was also referred to by several of his contemporaries, was used in harmonic theory by people who he himself referred to as the ancients.

Temple suggests that this apparent harmonic connection between Sirius B and the sun—stars that, on a universal scale, are virtually neighbors—might in fact be implicit throughout the universe, at least between localized white dwarf stars and ordinary stars like the sun. The wider implication is that all types of stars could have relative masses corresponding in some way to the established ratios of harmonic theory, that is, with the ratios embodied within the Hermetic Code.

One possible way of explaining this long-range coordination, says Temple, is to regard the two solar systems as inhabiting the same "cell" of space. This idea has emerged from a new area of research known as Complexity Theory, which involves the study of the sudden appearance and disappearance of order in the greater cosmos. It has been noted that something that looks very much like instantaneous communication occurs in such "cells," "whereby huge macro-regions of space behave as if their elements were not separated by spatial or temporal distance, and the 'cell' engages in what is called 'self-organisation.'"[6]

We have already identified what appears to be a microworld equivalent of this kind of process in plasmas, where billions of electrons simultaneously perform coordinated movements, exactly as if they were all communicating non-locally. Another example cited by Temple is the Bénard cell, a thermal phenomenon caused by convection in a fluid, in which millions of individual molecules instantaneously align. He also notes that there are other similar phenomena in nature, such as the simple sponge, which can transmit stimuli from one end of its body to the other at apparently "impossible" velocities, as if the whole creature were a single giant cell or neuron. This is not dissimilar to the proposition made by Roger Penrose, the Cambridge scientist mentioned in chapter 4, who suggested that "non-local quantum correlations" might occur between widely separated regions of the brain, thus enabling billions of individual neurons to respond as a coherent whole—a microcosmic equivalent of the Greek concept of homonoia, a "union of minds," or, in this case, of neurons.

Obviously the principle of nonlocality is hard for us to understand. It defies ordinary logic and excludes the time and space familiar to our ordinary senses. But while the nonlocal realm—what I have called the plane of light—might be difficult to conceptualize, there is a sense in which music itself can provide an explanation for the kind of simultaneous coordinated action we have been considering here.

This centers around the eighth and last note of an octave, Do, which, once struck, simultaneously becomes the first note of a higher octave, a greater scale. Such a note has dual properties, existing in two different scales at one and the same time. So let's say that the whole range of biochemical vibrations produced by neurons in the brain or in a sponge develop inwardly as an octave, and that ultimately this octave begins to vibrate, to resonate, at its optimum potential. When this occurs, the entire evolutionary scale becomes fused into one final note, Do. In this way all separate components of the scale not only become simultaneously interconnected with all other components, no matter what their "position" in the scale, they also become simultaneously connected, through the ultimate note, with the next scale or dimension above.

In the same way the RNA codon template, created by DNA from three inert chemical bases, or three harmonious "octaves" of chemical resonance, simultaneously becomes a single new biochemical "note"— an amino acid—one of twenty-two comprising the greater scale above. Thus, although the process is essentially linear, taking place in time, there comes a point where a kind of simultaneity definitely does occur, where lower scales are suddenly transcended, and where time and space count for nothing. The same could apply, of course, to the higher scales of biochemical evolution, perhaps when the amino-acid chain transmutes "up" into the scale of the protein macromolecule, or when the protein evolves up further into the scale of organs, or of glands, bone, tissue, and so on. In all of these transitional stages of evolution there must be points where the notes in one scale all combine to strike simultaneously a single new note up into a greater scale. Therefore, these "nonlocal correlations," in addition to being a general property of nature at the quantum level of existence, probably manifest at many different levels on the evolutionary ladder.

As Temple says, if a simple sponge can defy space and time at the bottom of the sea, then it is not unreasonable to suppose that these greater "cells" above can do so within the galaxy.

Inevitably, perhaps, Temple is ultimately drawn to consider the possibility that such macrocosmic cells, which he calls Anubis cells (Anubis being the jackal-headed deity of the Egyptian pantheon associated with the "dog star" Sirius), may be alive. "The vast Ordering Principle," he says, "may be an Entity."[7]

Quite so. Hermetic is genetic, and the musical symmetries evident in the planetary sphere of our solar system, and in the mass ratio of Sirius B and our sun, indicate that this life force may be prevalent throughout the entire universe. Remember also that the basic structure of all life-bearing phenomena is the spiral, the helix—and the entire cosmos, as we have seen, is positively teeming with these "serpents in the sky."

Our own solar system is comprised of nine such serpents, all coiled around the path of the sun, while the motion of the sun itself traces an infinitely greater helix winding around the central path of the galactic center.

The most distinctive of the nine "lesser serpents" described above is, of course, serpent Earth, from which has developed the evolving "solar mind" of the human race. Of course, if this greater helix is a cosmic "chromosome" developing in an organic fashion deep inside some kind of cell nucleus, then logically one would expect to find the greater body of the host cell all around it. In this case, the most obvious structure in evidence is that of the solar system itself.

Interestingly enough, when we look at the solar system in relation to the greater body of the Milky Way, its position appears strikingly similar to that of certain ordinary living cells. We could compare it, for example, to the position of a single blood cell in the human body. Like the solar system, a white corpuscle is structured around a central nucleus, or "sun." Floating around the nucleus are smaller components of varying size, complexity, and energy content, such as enzymes, mitochondria, ribosomes, RNA, and so on. These components all exist inside the body of the cell, floating around in a watery medium, a liquid membrane known as the cytoplasm. Beyond the walls of the individual blood cell and separating it from all others is more fluid membrane.

The boundary of the cosmic "cytoplasm" of the solar system might be defined as the sphere of the sun's immediate magnetic and gravitational influence, the sphere in which all the planets, asteroids, comets, and other orbiting materials are contained. The "cytoplasm," however, or the medium in which the components of the solar system exist and operate, would be infinitely more rarefied than the liquid membrane of the cell, or even the air we breathe on Earth, but it must be just as real nonetheless and it similarly must fill the whole system. Possibly this medium is light itself or, rather, the entire spectrum of electromagnetic radiation, which extends far beyond the boundaries of the solar system and is the medium in which all greater cosmic systems exist.

The spiral galaxy, consisting of billions of these "solar cells," is composed mainly of hydrogen-burning stars, vast, interstellar clouds of cosmic dust, or nebulae, and, one assumes, billions upon billions of planets, all whirling around a central nucleus of super-dense energy—a black hole, perhaps. And this great cosmic firework, with its immense spiral arms, as well as spinning around its central axis, is also hurtling through space at a velocity of around six hundred kilometers per second. Therefore, as with the solar helices, if we wish to perceive something of the galaxy's true form, we must try to visualize it not in the timescale involved in taking a frozen snapshot of it, a few seconds or a minute or so, but in that of the galactic being itself. In such a scale, a few "seconds" might be equivalent to hundreds of thousands or even millions of our years. So, if the four-dimensional structure of the galaxy could somehow be captured by time-lapse photography, after a few of its "seconds" or "minutes" we would see something very similar to the long body of the solar system described by Rodney Collin, an immensely elongated, shimmering spiral of electromagnetic radiation coiling toward infinity—another true helix.

This image of a long helical body really only describes the basic physical or four-dimensional form of one of these macrocosmic "chromosomes." But, like the DNA strand in a cell nucleus, the brain housed in a skull, or the creative solar mind of the human race, its overall complexity and influence would far exceed the scale of its origin. In fact, what we see when we look at the contents of the nucleus of a cell, a cerebral cortex or a solar system is merely a simplified cross-section of the whole entity.

For example, we look at a DNA strand and see only a relatively simple chain of chemically encoded digital instructions. Yet scientific investigation has shown us another, much more powerful dimension to DNA: it reaches out and indeed controls all of the creative functions in every part of the greater "world" in which it exists, that is, in the entire body of the host organism. Thus a single gene located in the chromosome of the first reproductive cells of an evolving organism may ultimately determine such features as the color of hair or eyes, the configuration of bone structure, and other complex characteristics.

Similarly the human brain can be scientifically reduced to its simplest form by describing it as a mass of neurons interacting through chemical reactions and electrical impulses, all comfortably housed in a protective covering of hard bone. But quite clearly the skull itself does not even remotely define the real boundaries of the brain's existence. The brain, like the chromosome, is merely the physical manifestation of a much greater, profoundly more complex entity, one capable of thinking conceptually or of dreaming up imaginary worlds, that of traveling backward in time through memory or alternatively speculating its way into the future. It can transmit information to other brains, it can intuit, impress, inspire, it can even, many believe, communicate telepathically, directly influence physical objects, predict coming events, and so on. In effect, like the DNA double helix, the brain is potentially as big as the "world" in which it functions.

The solar helix, or the collective mind of humanity, would clearly be of an order of consciousness far more advanced than any we could imagine. The "organism" in which this helix is housed would be composed of the entire body of mankind's accumulated wisdom, every idea, theory, or belief system that has ever been conceived, or ever will be. Trying to understand the true nature of such a being, whose life span would be measured in hundreds of thousands of our years, would involve studying, in the minutest detail, every intellectual and spiritual discipline known or yet to be developed. Like the two orders of helices below it—the human brain and DNA—we would expect the solar helix to exert creative influences reaching way beyond its own scale of existence, out into the greater body of the host galaxy.

In the same way we see the greater galactic helix as composed of symmetrical, localized concentrations of matter and energy traveling through a given region in space, but its greater presence, or the totality of vibrations issuing from it, spreads far and wide. We know the galaxy is formed like it is because there are four fundamental forces ("bases") keeping it together: the short-range strong and short-range weak nuclear forces, the gravitational force, and the electromagnetic force. The electromagnetic radiation emitted by all the stars of a galaxy spreads out at the speed of light in all directions, extending over distances of billions of light years from the source of origin. Therefore, the outer limits of all the light that has ever been emitted from the galactic helix, together with the outer limits of the gravitational influence it has exerted from the time of its formation, represent the greater body of the galaxy itself. Thus, as with the DNA double helix, the human brain, and the solar helix, we can say that the potential influence of the galactic "mind" would also be as immense and complex as the "world" in which it exists.

Scientists will argue that a galaxy cannot conceivably possess any kind of consciousness, that it is simply an involving, runaway mass of chemical elements randomly exploding and flying off in all directions according to basic physical laws. But then, the often-violent electrochemical reactions taking place inside an active human brain could also be described in much the same way—and yet we know that consciousness dwells there. Similarly the superactive speed-of-light fusion of electropositive elements combining, through photon interchange, with electronegative elements in the atomic chemistry of dynamic, evolving biomolecules would also give the appearance, to a microcosmic onlooker, of being a purely physical, entropic process. But we know that this entropy observed in the genetic microworld is basically an illusion, for from it evolve immense, harmoniously proportioned, and long-living organic structures.

In the case of the solar being, whose extraterrestrial body, remember, is constructed from the metaphysical "gene pool" of mankind's collective consciousness, disorder seems, at least on the surface of things, to be endemic. Go into any large town or city on a normal day and observe the inhabitants going about their business, rushing, pushing, shouting, hustling, absent-mindedly moving around in random directions, each of them in a private world of their own, with hardly ever a thought for the planet we live on, or the solar system within which it rotates, or the galaxy on high. No homonoia here. Elsewhere men are warring with and killing one another in a hundred different regions of the world, famines are ravaging millions of helpless and innocent victims with merciless regularity, global ecological disasters are occurring almost daily. All this evident confusion is "cacophony," a general manifestation of the social animal at its worst, with consciousness locked in a materialistic, dualistic stupor. No homonoia here either.

And yet beneath all this apparently chaotic activity there is, in fact, an underlying current of metaphysical harmony that has been continuously flowing throughout recorded history in the form of hermetic ideas. Fortunately for us, and presumably also for the Helix above, these concepts, being psychologically sound, are infinitely more "resonant" than the crass "isms" that man is prone to preach. This is precisely why, just like successful genes in the evolutionary processes of the microworld, they are so faithfully replicated and passed on for future generations by millions of other human minds.

This solar being, whose metaphysical "body" we have just described, is but one of around one hundred billion in our galaxy alone. If we assume that these beings possess "minds" with a degree of consciousness of some macrocosmic order, then presumably their "thoughts" or "concepts" would also have substance to them and would in turn be synthesized at a higher level in the construction of an infinitely greater galactic body. The real nature of such godly thought processes lie beyond our ordinary comprehension, but the manner in which they evolve must in principle be identical to the evolution of the helices below—the DNA molecule and the human brain. Therefore, the "concepts" or evolutionary signals engendered by the solar being above us, assuming they are of an "immaculate" order, would simultaneously be passed on, or transmitted, to other solar beings in the galaxy. So, like the dominant or active genes of DNA, or the hermetic ideas of creative mankind, the more successful of the "ideas" conceived in the solar helix will be replicated by other conscious beings in its "world." This will construct the body of an even greater organism—the galactic being.

But what about the mind of this greater entity? Where is it? How does it operate?

As I said earlier, the body of the galaxy is composed of billions of individual solar systems, or solar "cells," but its mind, evolving transcendentally out of the collective consciousness of all the solar beings in a given galaxy, is identifiable in the overall four-dimensional helical structure of the galaxy itself—the galactic "chromosome."

The order of consciousness of the solar being, as we have seen, is complex enough in itself, but it can, nevertheless, be explained in fairly rational terms, that is, as a composite structure formed from the entire body of humanity's collective consciousness, its accumulated secular and esoteric wisdom. But when considering the kind of "consciousness" our own Milky Way might possess, which clearly would exist and operate in a scale of being unimaginably greater than the solar scale, we are touching on possible processes so refined and ultra-resonant that they must remain for us hypothetical in our present state of evolution. This does not, however, prevent us from speculating on the nature of these projected "galactic vibrations."

Possibly the most distinctive features of these galactic vibrations would be their relative pitch frequencies and their rate of transmission. These higher creative processes would operate with degrees of resonance far more rarefied than those emanating from the helices below.

When we considered the solar helix, we identified its two principal properties or components as the active emanations of the sun (light) and the passive, metaphysical vibrations of the Earth (consciousness). These, I suggested, were "light" and "consciousness" of a different order from the light and consciousness of our ordinary world. Significantly the nature of the more rarefied light of the solar helix is described by the Hermetic Code and by the Magic Square associated with the Great Pyramid—"The Lights"—as a squared phenomenon, the square of the constant. Such, therefore, would be the nature of the "light" of the solar helix. The nature of the more rarefied consciousness of the entire human race would therefore have to correspond accordingly, and would presumably be as far removed from ordinary consciousness as the speed of light is from the square of the speed of light.

The galactic helix, however, whose scale of being is at least one hundred billion times more extensive that that of the single solar helix, would probably be engaged in an exchange of energies moving, or vibrating, at frequencies far in excess of the square of the speed of light. As we know, Special Relativity asserts that nothing can travel through space faster than the constant velocity. But of course we have seen from the nonlocal connections existing between interacting quanta that information can "travel" instantaneously from one to another—through the quantum field, where space and time simply don't exist. Possibly this is how these great galactic beings, whose sheer magnitude make even the square of the speed of light seem hopelessly inadequate as a universal rate of intelligence transmission, might "speak" with one another.

We are now poised to make a final ascent to the very summit of

Jacob's evolutionary "ladder," beyond the scale of the "angels" (suns) and the scale of the "archangels" (galaxies), and out into the realms of the Absolute scale—the universe in its entirety. In this scale, the mighty galaxy, whose three-dimensional form is measured in tens of billions of light years across, is but a single cell in the body of its host. And just like all cells, the cells of planetary organisms, the cells of the solar body (you and me), and the cells of the galactic body (like our solar system), this greater galactic mind must ultimately have the ability to create, in the greatest scale of them all, the ultimate, universal being.

Before we continue our journey across the universe, it is worth reflecting for a moment on the overall evolutionary picture we have just been describing. This is a picture, remember, that was first outlined by Egyptian metaphysicians in the third millennium BCE and that was neatly summed up in the phrase "As above, so below." Now you may or may not accept this scenario of ascending, living scales as the real thing, but whether this view is literally true or not, it is nevertheless unique in the entire history of philosophical thought in that it provides a very plausible answer to two of the most fundamental and puzzling questions of all, questions that, as I explained in the introduction of this book, were the cause of much consternation to me as a boy: why are we here? Is there life after death? The theory of transcendental evolution pulls no punches here: it answers these two questions in a quite straightforward and unambiguous way.

According to this original creation theory, we are here as a direct result of nature's grand design, all of us being—potentially at least—vital and integral parts of a much greater evolutionary process. This process begins in the "primordial waters" (with the DNA–RNA complex), it then evolves up through the consciousness of sentient beings like ourselves, then further still into "angelic" (solar) and "archangelic" (galactic) form, ultimately to flower into the superconscious "mind" of the universe itself—the ultimate "helix." We are a crucial link in the chain.

On the second question—of life after death—hermetic theory is equally emphatic. Of course there is life after death, for death itself, the final note, Do, at the top of one's own personal scale of evolution, is also the first note of the greater scale above. By this account, not only is there life after death, but, compared to the ordinary timescale of the modern hominid, it would be, as the ancients have always said, a "life everlasting."

The implication of this upward evolutionary motion is that the universe seems destined to become fully conscious of itself. But then perhaps it is already; it is certainly old enough to have come of age by now. Maybe this is why photons are so acutely aware of happenings in the greater quantum field, or why billions of electrons in plasmas and metals can act as if they already know what billions of other electrons are about to do. It's as if there is a general "awareness," even at the most basic level of material existence.

Sri Aurobindo said that the universe was wholly conscious and that if just one point in it were not so, the whole fabric would break down into a lifeless void. He was merely echoing the Greeks, of course, but the message remains the same, which is that the universe is, in fact, already conscious of itself, and that it is merely waiting for us to realize this and contribute toward its maintenance. Perhaps this is why the God of the ancients was said to be so concerned for our well-being. We are his life-blood.

9
The Hermetic Universe of Ancient Times

By now readers might appreciate how important and significant are the evolutionary ideas of the ancients, and in particular the musical revelations of the Pythagoreans. We know, however, that Pythagoras, like all other great spiritual leaders, was merely passing on knowledge that came originally from the priest-astronomers of ancient Egypt. Possibly the Egyptians also inherited the main tenets of this wisdom from the fabled flood survivors of ancient myth, who themselves could possibly have received instruction from an even earlier race. This continuous evolutionary line appears to have originated in the belief system of the "primitive" Neanderthal, who regarded the number 7, the fundamental symbol of the octave, as sacred. The seven bear skulls found in the stone altar at the Neanderthal site at Drachenloch in Switzerland indicate that this sacred number symbolism dates back at least 75,000 years.

With the Greeks, however, came a much more overt, logical description of the theory of transcendental evolution, which the Pythagoreans neatly summed up in the two key esoteric symbols already discussed, namely the classical formula *pi,* 22/7, and the original "philosopher's stone," the Tetrad, illustrated by placing ten pebbles on the ground in the shape of a 4–3–2–1 triangle.

The Tetrad was called by the Pythagoreans the "model of the gods" and the "source of nature." It was thus regarded as the blueprint for the development of all evolutionary phenomena, above and below: the 4–3–2–1 format of the symbol is in fact a remarkably accurate blueprint of the processes involved at the biomolecular level, for it describes perfectly the sequences of genetic processes involved in the synthesis of amino acids.

The formula *pi* is also an expression of the same evolutionary process. So the four-base/triple-octave symmetry embodied in the classical convention, 22/7, which can be expressed diagrammatically like this:

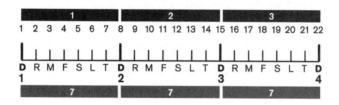

denotes the first two levels of the Tetrad:

The two higher stages in its evolutionary development, marked by the two pebbles at the third level and the single one at the apex, are a combined expression of the greater "trinity" above.

When this model is applied to the higher evolution of the individual, the combined four- and three-pebble stages, with their four-base/triple-octave symmetry, represent the fundamental qualities of all human beings—walking trinities with the capacity to sense, emote, and perceive. By and large, we can all do these things to greater or lesser degrees; they are perfectly natural human functions. The next two stages in the Tetrad, however—the third, denoted by the two pebbles, and the fourth, with its single pebble at the apex—stand for higher human functions that unfortunately are not universal. This is where what we might call "original thought" comes into play, which is the harmonious product of a balanced combination of our sensations, emotions, and perceptions. This spark of real consciousness is denoted by the first of the two pebbles at the third stage of our evolutionary Tetrad. The second pebble represents the other side of this metaphysical coin: light itself. The topmost pebble therefore symbolizes the final, transcendental note of this whole musical process. Generally referred to nowadays as a "concept," this signal, harmonious and therefore transcendental, then continues to exist as a single new note in the greater scale above. We see from the genetic code that the "greater scale," the scale up from the base scale of the four chemical bases, consists of precisely twenty-two higher notes—twenty amino-acid signals and the two signals coding for "start" and "stop"—a triple octave. It follows, therefore, that the greater scale into which the conscious mind can input is also structured as a "triple octave," a "trinity" above.

We can now try to apply the process described by the Tetrad to the greater cosmic scales outlined in the last two chapters. If solar beings and galactic beings are for real, they should fit easily into an overall hermetic picture of universal events.

As we have noted, the Greeks believed that all cosmological entities like the planets and the stars were conscious beings, and that the universe itself was a living animal—a zoon—and therefore completely organic in nature. We have seen how such an organism might develop, through an ascending hierarchy of scales, from biomolecules to galaxies. Remembering that there are exactly four of these fundamental scales, we can envisage the whole universe as being a vast, cosmological representation of the Hermetic Code itself, a multidimensional Tetrad:

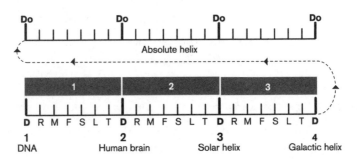

One of the most significant features of this diagram is that the four orders of intelligence depicted, from DNA to the galactic helix, are each represented by the note Do. That is, they are all manifestations of the very same note; only the scale is different. This, of course, is precisely what is being alluded to in that all-embracing dictum of Thoth, "As above, so below," which tells us that the symmetries of the processes of creation are the same at every level of existence, above us, below us, and in between, of course, in our minds.

So these four basic orders of intelligence or life all resonate at compatible frequencies, with each successive note, Do, vibrating, according to musical theory, at exactly twice the pitch and frequency of the preceding one. An octave, remember, is a measure of the doubling of the rate of vibrations in a given scale. This indicates that there are unique, tangible connections between the four "base notes," the vibrations of each being whole-number coordinates of the greater evolutionary scale.

Theoretically the super-resonant galactic helix, representing the ultimate note, Do, of the third and final evolutionary octave, would ultimately have the power to strike a single new note up onto the greatest scale of them all: the universal scale. What happens beyond that is anybody's guess. I have found myself trying to envisage here a dynamic, cyclic scenario, whereby a given proportion of the energies created by the galactic helix reenters, possibly through the quantum field, the primary DNA scale. After all, some thing, some kind of force or intelligence, is ensuring that the universe manifests and evolves strictly according to the laws described by the Hermetic Code, and the most obvious choice as to the possible source of this intelligence surely must be the ultimate product of the whole evolutionary process. Remember that the galactic helix, the fourth and last base note of our universal triple octave, is in fact reinforcing, at a higher pitch and frequency, the first base note of the entire scale, represented by DNA. But DNA is not simply a note. It is also an entire scale, and the first "note" of this primary DNA scale would have to be one of the base notes of the genetic code, one of the four inert chemical bases. Conceivably therefore, it could be at this stage, on the level of the simple inorganic molecule, that the creative vibrations of the galaxy above reenter, through the quantum field, the endless cycle of life. Thus the chemical base might seem inert from a scientific perspective, but in reality it may have already been imbued by the powers above with some sort of rudimentary, radarlike intelligence, providing it with at least enough awareness to be in the right place—the living cell—at exactly the right time.

REINCARNATION

As an interesting aside, it is worth pointing out that the process of evolution described above hints at a possible explanation for the emergence of the Buddhist and Hindu beliefs about reincarnation, which is also a cyclic description of evolution.

The Pythagorean concept of metempsychosis, or the "transmigration of souls," expresses much the same idea. Pythagoras regarded the soul as a fallen angel locked within a body and condemned to a cycle of rebirths until it has rid itself of all impurities. The cycle being described, from birth to death to birth again and so on, could be regarded as being, in a sense, circular, where the evolving entity keeps returning back to the point of departure, or to the moment of its conception. But then, if the soul were improving its lot at each turn, this would imply a slight "upward" movement after each cycle, one lifetime being superimposed on top of the next in ever-ascending circles. This is significant, for, if we were to draw an imaginary line tracing the path of this recurring entity as it gradually evolved, the overall figure so described would take the form of the most fundamental configuration of all evolutionary processes—a helix.

Unfortunately this question of recurring lifetimes represents something of a departure from the main thrust of this study. A detailed investigation would require a great deal more time and space than I currently have. Possibly some time in the future we might be able to investigate this subject in more detail, but for the moment we shall continue our search for evidence in support of the ancients' view of a living cosmos.

THE CONCEPTION OF THE UNIVERSE

The Greeks' definition of the universe—a zoon—is wholly unambiguous. They regarded the whole cosmos as the biological product of a fertilized ovum, a living, organic creature conceived through some form of procreative activity. By whom or in what is clearly the most profound mystery of all.

Whatever its genealogy, however, most origin myths agree that the present universe was created, or rather conceived. Take the example most familiar to Christians: "In the beginning God created the heaven and the earth."[1]

In Genesis, the creation or conception of the universe is described as having taken place in a watery medium, which in ancient scriptures always has a feminine or passive connotation: "And the spirit of God moved upon the face of the waters."[2]

Then comes the moment of conception, the initial act of (pro)creation: "And God said, 'Let there be light'; and there was light. . . . And God divided the light from the darkness."[3]

So the primordial cosmic "ovum" divided into two complementary yet quite distinct proto-cells, one light, one dark, or one active, one passive.

In a similar vein, the Vedic version of universal origins asserts that God "first with a thought created the waters, and placed his seed in them."[4] This again suggests that the origin of the universe was primarily a natural biological event.

In Vedic literature there are hymns dedicated to the god of the primeval waters. This is Indra, the god of rain, who is said to have released the waters to flow into the cosmic ocean and to have revealed the creative light of the god Agni—the sun.

To the early Greeks too, water was considered a primary element of creation. The philosopher Thales, for example, believed that the Earth floated on water, which was the medium from which all life evolved. Much the same view was held by the Pythagoreans, who thought that sunlight penetrated the primeval slime of the Earth to generate life.

Of all known origin myths, the Egyptian account is possibly the oldest. Thus each of the above examples is merely a reprise of the original theme, first set out by the priests of Hermopolis, the spiritual seat of Thoth. Hermopolitan myth speaks of eight principal gods who appeared simultaneously on the "Island of Flame," which rose like a hill from the eternal waters.

As we can see, virtually all of these creation myths agree on two fundamental points: first, that before the universe/world/life came into existence, there were only endless or eternal waters—the passive, negative element—and second, that the creative act itself involved the introduction of light, or a flame—the active, positive element. Very often this fusion of forces is described as having occurred through the intervention of a god or gods—the universal mediating principle. Excluding this latter allusion to "divine intervention," we are left with a description of the universe's creation that in fact bears striking similarities to that currently on offer from modern science.

THE SCIENTIFIC PERSPECTIVE

Possibly many readers will already be familiar with the "big bang" theory of the origin of the universe, a proposition first put forward by the Belgian priest-astronomer Georges Henri Lemaitre in the 1920s. This is now generally accepted as the most likely explanation of how matter, space, and time came into being. A persistent background microwave radiation spreading out evenly across the entire cosmos and with a temperature of around 3.5 Kelvin (3.5 degrees above absolute zero) was recorded by the radio-astronomers Robert Wilson and Arno Penzias at Bell Laboratories in 1964. Most scientists now agree that this radiation is very probably the residual vibration of the creation of the universe, of the biggest bang in history.

But what exactly was it that originally went bang? Lemaitre suggested that the universe had been born from a single primeval quantum of potential energy, a kind of superdense mother of all atoms. After the initial cataclysmic explosion, this primordial "atom" began dividing so rapidly and energetically that it eventually gave rise to all the matter in the universe. As the first atomic nuclei (protons and neutrons composed of quarks) proliferated, with quantum duplication taking place at a phenomenal rate, space and time simultaneously unfolded to accommodate them. This means that before the primordial quantum split asunder and the resultant superhigh energies began to radiate out from the "epicenter," there was no space, no time, nothing except the original quantum itself.

Lemaitre realized that quantum theory supported this idea of space and time appearing after the big bang. As we saw earlier, in quantum mechanical calculations, space and time are statistically meaningless in respect to individual events involving subatomic quanta. Therefore, if the universe did originate from a single, self-duplicating quantum, space and time would not have existed at that point; they would not have appeared until the primordial "atom" had duplicated in sufficient quantities to produce a significant number of measurable quanta.

According to big-bang theorists, the universe was in thermal equilibrium during its earliest development and was filled with the most intense light traveling out in all directions ("And God said, 'Let there be light'"). The temperatures involved at this stage would have been in the trillions of degrees. The original wavelength of these first generations of photons would have been very short, but as space expanded it stretched out the wavelength of the light, so producing a one-way shift to lower and lower temperatures—white light shifting to blue, blue to red, and so on. The present cool state of the universe, barely 4 degrees Celsius above absolute zero, is the end result of this fifteen-billion-year-long fireworks display.

Within half a billion or so years after the primordial conception the force of gravity caused pockets of high-density dust clouds and atomic nuclei to condense into galaxy formations. We can still observe such a process at work in the creation of proto-stars ("baby" stars) forming as dense clouds of cosmic dust collapse inward, such as is currently being observed in the Large Magellanic Cloud system, a member of our own immediate cluster of local galaxies.

Individual stars within these galaxies are all born as proto-stars. As they develop through high-energy nucleon collision caused by gravitational collapse, these baby stars rapidly approach maturity and ultimately "ignite," converting hydrogen to helium at a phenomenal rate. At this stage they are classed as mature, "main sequence" stars—like our own sun in its present state. Main sequence stars, after billions of years of relatively constant, active life, eventually metamorphose into old-timers—red giants. Red giants then either degenerate gradually to become static white or brown dwarfs, or they reach a critical energy level and explode as supernovae. A supernova is a star that has become pregnant with a vast store of nuclear energy and ultimately explodes, projecting massive quantities of radiation and heavier chemical elements back out into the cosmos, where it is then recycled. It's an interesting reflection that every single atom of which you and I are composed came from exploding supernovae out there in deepest space.

Until very recently it was thought that any region of space was much the same as any other—that galaxies developed relatively undisturbed by other concentrations of mass. This view of a uniform distribution of galaxies was initially supported by data obtained from high-altitude flight experiments using redeployed U2 spy-planes. These experiments, coordinated by the American astrophysicist George Smoot in 1995–96, appeared at first to show that the universe is expanding uniformly and with a constant speed in all directions. However, more accurate experimental procedures later revealed that this was not so and that in fact galaxy densities are not strictly homogeneous and that there are huge clusters of galaxies gathering in some regions and vast expanses of empty space in others.

Our own galaxy is a member of a relatively small local cluster, all hurtling through space at a velocity of around six hundred kilometers per second. Current theory holds that the extraordinarily rapid motion of these massive bodies is caused by the gravitational pull of a very large concentration of mass situated a great distance away. This "Great Attractor," as it is called, is thought to be another, incredibly vast cluster of galaxies, a kind of supercluster situated millions of light years distant. These greater galactic "cluster cells," varying so dramatically in size and luminosity, indicate that the expanding universe is far from symmetrical, that its "body," like yours, is lumpy and uneven and much more structured than had previously been thought.

Astrophysicists have now discovered the "seeds" of these structural characteristics in slight fluctuations in the cosmic background radiation, which suggests that they must already have been present in the fabric of the universe as little as 300,000 years after the big bang. These early seeds were the primordial imprints of creation, "cosmic genes" in which were encoded all the characteristics of the universe as it exists today.

Science currently recognizes four fundamental forces in the universe: the weak and strong nuclear forces, the electromagnetic force, and the gravitational force. An instant after the big bang, however, there was only one unified force: matter was indistinguishable from energy, and the first rudimentary quanta—the quarks—had not yet been formed. These high-energy conditions at the very beginning of time are now the focus of much attention. Scientists believe that a fuller understanding of the nature of universal origins will come through a rational convergence on the first moments after this unique moment of "conception," when only one unified force existed and where the laws and the components of the universe were much simpler than they are today.

In his book *Wrinkles in Time,* George Smoot uses an interesting analogy to describe the nature of his work. He compares the quest to understand the origin of the universe by converging on the moment of creation to that of tracing the evolutionary development of the human being back to his or her origins. The human being is an immensely complex entity with definite and unique physiological, emotional, and psychological characteristics. But if we trace such an entity back through its life toward the moment of its conception, it appears progressively simpler in structure, until ultimately we find a uniform set of relatively simple digital instructions encoded within the chromosomes.

Smoot is obviously using this comparison between the universe and the individual only as an analogy, but, like so many cosmologists and astrophysicists today, he seems particularly fond of biological metaphors. For example, he says that "the universe appears to be as it is because it must be that way; its evolution was written in its beginnings—in its cosmic DNA, if you will."[5] He also talks of "quantum self-replication" taking place at an explosive rate very soon after the primordial event, much the same as individual cells self-replicate at an "explosive" rate as a living organism rapidly "expands" after conception.

Another example of the use of "bio-cosmic" metaphor is given by Professor Paul Davies in his book *The Last Three Minutes,* in which he discusses a proposition made by a group of Japanese physicists working on the idea of "false" and "true" vacuums. A false vacuum is an excited vacuum, a region of so-called empty space in which a great deal of quantum activity (particle interaction) is still present. The natural tendency of a false vacuum is to decay to its lowest possible energy state—a true vacuum. The Japanese postulated an alternative process based on a simple mathematical model, where a small bubble of false vacuum surrounded by a true vacuum would inflate and subsequently expand into a larger universe in a big bang. Davies uses the analogy of a rubber sheet (representing the true vacuum of an existing universe) blistering up in a given place and ballooning out to form a "baby universe," connected to the original universe by a "wormhole," the opening of which would appear to an observer in the mother universe as a black hole. The black hole then evaporates and finally disappears, pinching off the "umbilical cord"—the wormhole—leaving the baby universe, a high-energy false vacuum, to grow and develop independently.

Here again we have a scientist using what appears in recent times to have become the accepted idiom for describing cosmological processes—the biological metaphor. Popular books on cosmology and astrophysics now abound with such terms, and one begins to wonder whether this is simply a fashionable trend, or is it, perhaps, some deeper influence affecting the development of human consciousness.

We touched earlier upon the possible nature of this influence, when I proposed that human ideas or inventions could be regarded as the metaphysical equivalent of the amino acid, or perhaps a chain of amino acids. A string of related ideas, which together make up what we would call a full-blown concept (such as the Hermetic Code, for example), we might call a metaphysical "gene," or perhaps a chain of genes. Now genes can be either "dominant" or "recessive," active or passive. They can lie dormant in the human genome for generations and they can reemerge once more as dominant genes anytime conditions become favorable.

Perhaps this is what is happening now in respect of the Hermetic Code. It is surfacing once again, and while science has been systematically proving the existence of hermetic symmetries at all levels of material and biological creation, simultaneously there has been a great upsurge in awareness of the remarkable achievements and beliefs of our remote ancestors. Remember, the Hermetic Code has been the dominant feature of human consciousness many times before, in the time of Muhammad, for example, and of Jesus, Zoroaster, Pythagoras, Buddha, Confucius, Moses—the list goes on and on, back into the mists of time. It is entirely possible, therefore, that we are currently witnessing—participating in, even—the beginnings of yet another renaissance in the development of human consciousness, the emergence of a new, "modern" version of the oldest creed on Earth, one that naturally requires us, either consciously or unconsciously, to reinvent the hermetic universe.

Arguably the best example of the recycled concept currently on offer is the theory of universal origins proposed by the physicist Lee

Smolin. Smolin has suggested that there may be a kind of Darwinian natural selection taking place among universes and that the emergence of organic life and conscious beings is a by-product of this process. In other words, he is proposing that the universe is a zoon.

Clearly this "natural" conclusion is just about as close to the process I am trying to envisage as it is possible to come, for not only does it agree with the known scientific facts concerning the origin of the universe, it also happens to fit all the criteria of the hermetic view of creation.

We earlier noted Smoot's discovery that galaxies, like stars, are grouped in clusters—cluster cells—and even superclusters. This gives a universal structure and pattern of development very reminiscent of the way living cells gather together in clusters to create a variety of organs, bone, muscle, nerve tissue, skin, and so forth. So perhaps the Great Attractor, the immense supercluster toward which our local group is surging, is an "organ" of some kind in the body of some great being: its "heart," an "eye," or even its "brain." If this were the case, then the relatively small local cluster of galactic life forms, on the back of one of which we are presently riding, might seem lowly and insignificant, but like, say, a blood cell entering into a vital organ of the body, our galaxy would be a contributor to life itself.

THE METAPHYSICAL PERSPECTIVE

Many readers will probably be aware that this hermetic picture of an evolving, organic cosmos is completely at odds with the orthodox scientific version of events, which holds that the universe is essentially an involutionary phenomenon and that, given enough time, all physical systems within it must ultimately descend into chaos. The basis of this assumption is the most fundamental scientific law, the second law of thermodynamics, which says that energy has a natural and irreversible tendency to dissipate. This is what is apparently happening in the universe all the time, where high-density pockets of energy are unevenly distributed, mainly in stars, but also in planets and interstellar space. All this energy is continually dispersing, and on our own planet this is what provides the impetus for all the chemical reactions that make life possible.

Unlike closed physical systems, which simply "waste" their energy, biological systems are highly organized entities, continually evolving into states of ever-increasing complexity. They are intelligent, in tune with their environment, and so are capable of "exporting" entropy (disorder, chaos) and of bringing in energy from outside themselves to sustain their own regenerative and creative processes. As a cell grows and ultimately self-replicates, it is continually taking in energy from its environment and using it to manufacture essential biomolecular components. Similarly we ourselves take in "free energy" in the form of food, air, impressions, light quanta, and so on—all of which are residual products of the greater, entropic movement of a thermodynamic universe. Thus, say scientists, organic systems do not actually violate the law of thermodynamics; they are simply able to temporarily evade the overall degenerative process as and when physical conditions are favorable. So we are all, in a sense, living on borrowed time. When the primary source of our energy—the sun—begins its inexorable descent into chaos, life on Earth will become history. Life in time, that is.

But what about the proposed higher forms of "life" discussed earlier? What about all the solar beings in all the galaxies and all the galactic life forms existing throughout the entire universe? Surely such entities, once created, would continue to exist and to evolve over billions of years irrespective of the dissipative physical energies harnessed in a given, isolated planetary system. Thus the heart of the solar cell—its sun—may die, but its "higher self," or the creative "genes" synthesized during its lifetime, must live on in the greater galactic scale. We earlier ascertained that solar and galactic helices, if they are a reality, would exist in other, greater dimensions—on the plane of light, for example, or in the quantum field—where there is no time as we know it and therefore no frame of reference within which to define a degenerative dispersal of energy, an increase in entropy. This would explain why a photon can travel across the entire universe and still maintain the maximum velocity possible—because at the speed of light it is free from the ordinary ravages of time.

Clearly, therefore, there could be processes in the universe that continue to unfold irrespective of the directional flow of time. What is more, if these higher organic life forms do indeed exist, and all solar and galactic systems are by and large becoming more and more "conscious," then we might say that the overwhelming tendency of the greater universe is to become less and less "chaotic" as it evolves.

In *The Infinite Harmony* I suggested that the human animal, composed of billions upon billions of cells, is, in effect, a universe in miniature, whose highly organized structures and functions are created from the coordinated activity of a host of chromosomes, or microcosmic "galaxies." Such a body is conceived and then born, after which it grows through successive stages of development until it reaches maturity. Ultimately it gives up the ghost and subsequently releases its component particles, through natural decay, back into the entropic void. It is, however, possible for the human being's emotional, psychological, and spiritual output to continue long after the body has passed its prime and begun its inexorable descent back into the ocean of chaos. Furthermore, even when a given individual is defined as "dead," though virtually no trace of his or her physical existence remains in space and time, the overall influences generated during his or her planetary existence—ideas, impressions, concepts, and so on—can persist, as in the well-documented cases of history's major religious figures, for millennia. In a sense, these influences exist independently of the ordinary time of the individual, whose life span is measured only in decades.

Obviously, therefore, if the universe is alive, then presumably what is being observed through the eyes of astronomers and astrophysicists represents only its physical body developing in time. Its higher conscious functions, that is its "emotional," "psychological," and "spiritual" worlds, would be invisible to us, ostensibly because such processes would be operating in spheres that reach way beyond the boundaries of the physical body, in the realms of the other realities already discussed, in which statistical notions of space and time lose all meaning.

These "spheres" and their respective boundaries are the subject of the next chapter. We have already divided the cosmos into four fundamental scales or orders of "intelligence": DNA, the human brain, the solar helix, and the galactic helix. But it is possible further to integrate these four scales into a more comprehensive cosmic picture by considering them in respect of another essentially hermetic concept, based on the assumption that the hermetic universe, a four-centered, living entity, exists and operates within an overall framework of seven interpenetrating dimensions.

10
The Hierarchy of Dimensions

While the "organic" universe is constructed from four basic orders of "intelligence," hermetic theory tells us that this creature must exist within the framework of an octave, that is, of seven, or even eight, dimensions. Most people recognize only the three dimensions of space and perhaps the fourth dimension of time, a greater "line" along which everything moves, as it were, in the direction of eternity. But the true picture, as we shall see, may be much wider in perspective, much more holistic than the reality we ordinarily perceive.

Let's start at the beginning, with the zero dimension, which in geometry would be defined as a finite point. If this point were to move in any direction, it would trace a line. A line is a one-dimensional entity and can be defined by its length only. Two dimensions would unfold if, for example, the whole line were to move in a sideways motion, so tracing a plane, having both length and breadth. Similarly, three dimensions would be described if the whole plane were to move in any direction at an angle to its surface, thus tracing a solid, with length, breadth, and height. We ourselves, at our most basic level, are three-dimensional entities, and so are the familiar sense-objects that make up our world.

As we see, a greater dimension unfolds every time a new direction is described. A moving point describes a line, a moving line describes a plane, a moving plane describes a three-dimensional solid.

Now all three-dimensional objects are also, in a sense, moving in another quite different direction. They are all getting progressively older, they are all existing along their line of time, their fourth dimension. This is the highest dimension that can be perceived in our ordinary states of consciousness. Hermetic theory, however, calls for at least seven of these expanding spheres, so in order to identify these otherworlds, we obviously need to stretch our imaginations somewhat and reach out beyond the realm of sense experience. Physicists have already paved the way in their attempts to conceptualize the next dimension up from the line of time. This is the curious realm of the quantum, the nonlocal arena of inner space existing beyond ordinary time. We can describe it in simplified terms as the dimension that would unfold if the line of time were somehow to move in a direction perpendicular to itself, so tracing a greater "plane." This is the fifth dimension, the "plane of light" discussed in earlier chapters.

We have seen how the first three spatial dimensions describe a line, a plane, and a solid. And if, as hermetic theory says, "above" is intrinsically the same as "below," with a difference only in scale, then we should expect a similar relationship to exist between the higher dimensions. Therefore, if the fourth dimension of time is a "line" and the fifth, the nonlocal sphere, is a "plane," then the boundaries of the sixth would define what we might call the "solid" form of the ultimate reality.

One might assume that this cosmic hierarchy of dimensions must end with the sixth, but we have already established that, if the cosmos is hermetic, it must be structured as a fundamental octave, so one would consequently expect the hierarchy of dimensions to reflect this order. We can therefore make one final conceptual leap by positing a seventh sphere, which could be defined as the medium in which the whole universal phenomenon exists. Paradoxically, however, this seventh dimension could lead us right back to the very (zero) point from which we started, for in such a reality, even the "medium" in which the universe exists (its street, city, planet, or whatever else might constitute its "space") might simply be the equivalent of a finite point in an unimaginably greater sphere.

We thus have seven interpenetrating dimensions coiling one out of another in ever-increasing spheres, beginning with a point and ending on a point. If we now remember that an octave also begins and ends on the same note, we can see that the hierarchy of dimensions fits in perfectly with the hermetic description of the universe.

Now that we have a relatively ordered picture of this seven-dimensional "ladder," we can try to ascertain our position within it. I hope readers will find this at the very least an interesting intellectual exercise and, at best, perhaps a way of understanding that, hidden deep within our nature, we human beings do in fact have a deep and profound affinity with the wider universe.

We can begin with one of the basic premises of hermetic theory, which says that what is above is the same as that which is below. Taken quite literally, this means that all of us are microcosmic copies of the universe itself—"images of God."

If this is so, then this miniature universe of "galactic" or chromosomal life forms must exist within the framework of seven dimensions, the equivalent of three spatial dimensions, one of time, and three more ascending spheres, corresponding to the "plane of light," the "solid of reality," and, finally, the inexpressible seventh dimension, the "medium" in which the whole exists.

Now if chromosomal DNA is the microcosmic equivalent of the double helix of the mind, we can say that, like the brain, it is housed in a three-dimensional structure living in its own dimension of time and that there must be other dimensions existing above and below it. Below the three- and four-dimensional scale of DNA, an intelligent, organic molecule, we have the scale of the much smaller inorganic molecule. The difference between inorganic and organic is vast. They are literally a dimension apart, and so we can consider the inorganic molecule as a relative manifestation of a two-dimensional plane. Moving on down, we come to the atomic scale, the equivalent, perhaps, of a one-dimensional line. Finally we have the chromodynamic scale of the electron and other subatomic waves and particles—points in space.

We now come to the dimensions above these chromosomal life forms, the dimensions existing beyond their space and time.

To identify these we need first to consider the overall lifetime of this miniature universe—that is, the human being—and the huge developmental leap from the DNA double helix to the double helix of the mind. All of this takes place in time, at least from the atomic scale upward. (Subatomic quanta, remember, exist in a timeless, nonlocal, zero dimension.)

Obviously DNA's scale of time is vastly more compacted than the timescale of the conscious human being. The cell is born, it works frantically all its life, and then it dies, or rather divides, in a matter of hours, days, or weeks. But of course its influences—its genes—live on through the chromosomes, endlessly dividing and multiplying for several decades. If the single cell could have any conception of its own time and, like us, speculate beyond its own experiential existence, several decades would seem to it like an eternity. And if some form of superior microcosmic intelligence were to suggest to the cell that its "soul," after death, or division, would in fact live for eternity, this humble little grafter might find such a notion a shade fanciful. And yet, this is precisely what does happen. The cell's influences, its genes, continue to be passed on through millions and billions of generations of other cells until the greater organism—its "universal host"—ultimately expires. Thus we might say that the body of the host organism not only represents a higher dimension for the cell, it is also one into which the cell can actively input evolutionary data. Let's call this dimension the chromosome's equivalent of the plane of light, the timeless, "eternal" fifth dimension.

The sixth dimension of our miniature universe, like the "solid" form of the ultimate reality described earlier, must be of an order infinitely greater and more complex than the fifth, planelike sphere—that is, the greater physical body inhabited by all of the organism's cells. It would be a dimension that would unfold if an entirely new direction were taken, that is, if all the cells in the body of a living organism were to combine and expand in some way, as a line expands into a plane and a plane into a solid. This, I would suggest, is where the consciousness of the organism kicks in, where the double helix of the mind, the ultimate creation of DNA, is finally formed. Clearly creation of the conscious mind of this miniature universe is a genuine transcendental phenomenon, existing and evolving in an infinitely higher scale of being to that of the single cell. This would be the cell's sixth dimension. To this higher conscious mind, the "eternity" of the individual cell, or the sum of all the lives of all of its body's cells, is perceived as but a single lifetime. What is more, the mind of this greater being, the cell's "god," possesses self-awareness and is fully conscious, not only of its own physical existence (which is the entire universe to the cell), but also of its environment, of its world at large and, perhaps, of other beings similar to it. From the perspective of the cell, therefore, this inexpressible environment, the home of its microcosmic universe, would be the seventh-dimensional medium in which its sixth-dimensional god exists.

As we see, the seven dimensions unfolding in the biomolecular world are related in the same way as the greater universal framework of dimensions described earlier: point to line, line to plane, plane to solid, then further on to a greater "line," a greater "plane," a greater "solid," and, finally, a "medium" in which the whole exists.

If we now apply the DNA model to the greater scale of existence of the human brain, then we can say that the successful "genes" created by the double helix of the mind—its ideas or concepts—exactly like the genes of the individual chromosome, can in fact last for "eternity"; that is, they can exist in the fifth dimension, on the timeless plane of light, or within the collective consciousness of the entire human race. Genes can do that; they can permeate through to every cell in the body. Likewise, objective concepts can do exactly the same thing; they can permeate through to every other conscious mind on the planet. More importantly, however, if the theory of transcendental evolution holds true, such concepts would also in the process be actively contributing toward the creation of an infinitely greater, universal consciousness.

So "God" is a six-dimensional entity. And so, in a very real sense, are you; only the scale is different. But, of course, this mighty macrocosmic being would be six-dimensional only to us. From the perspective of the DNA strand or that of the individual cell, our "God" would represent six dimensions squared, which means that the "medium" in which the universe exists would represent seven dimensions squared. Add to this the eighth—point zero—and we arrive at our now familiar hermetic concept, which holds that the ultimate creative element is the product of the square of the constant.

So far we have identified two coexistent and interpenetrating "universes," the one in which the cell exists and the greater universe in which we ourselves exist. DNA forms the nucleus of a cell in the body of its six-dimensional "universe"—the human being—and the double helix of the mind is the nucleus of a cell in the body of a greater six-dimensional universe existing as some godlike being of inexpressible form and character.

Now, just as the double helix of the mind evolved from the cumulative work of hundreds of billions of cells and chromosomes, then we would expect a similar process to develop in the next scale of evolution, the scale of the solar helix already posited. So as the solar body grows out of our concepts, a higher six-dimensional "mind" should eventually evolve from it. This would be the "mind" of our perceived "universe," our God. But this solar mind, like DNA and the human mind, must also be a "chromosome," a creative, organic intelligence existing in the nucleus of a cell in the body of its "universe," in the next scale up, the scale of the galactic helix. And again, if the whole is developing hermetically, then ultimately an even greater "six-dimensional" galactic "mind" would evolve, presumably functioning as a "chromosome" in the greatest scale of them all—the "body" of its universe.

Thus each scale has its own "universe," and they are all inextricably interconnected, each being six-dimensional in relation to the one below it, each living and evolving in a seven-dimensional arena.

TIME: A ONE-WAY TICKET?

As we have noted, it is possible to differentiate between dimensions in terms of their relative times. The individual cell's time, for example, is very much more compacted than the time of the human being. The "eternity" of the cell, or the sum total of all the lives of all the cells within organisms like you or me, is equivalent to a normal human lifetime. Similarly the time of the human being must be equally compacted in relation to the timescale of the solar being, to whom our "eternity" would likewise be perceived as but a single lifetime. By the same token, the lifetimes of all solar beings—their "eternity"—would be a single lifetime to the galactic being, whose own "eternity," the sum of the lifetimes of all galaxies everywhere, would in turn represent the lifetime of its god, the ultimate universal entity.

Time, therefore, is variable, relative. But what is it exactly? Is it something that flows like a metaphysical river, gathering up everything in its wake? Or is the whole phenomenon, as mystics and shaman have always believed, simply an illusion? According to the experimentally verifiable theories of modern physics, of course, the shaman and priests have been right all along; time doesn't really exist. If you were able to travel at the speed of light, the "river" of time would apparently cease to flow. At least, that's how the scientist sees it. But hermetic theory hints at another possible scenario. It suggests, in fact, that this state of timelessness described by the physicist is also an illusion of sorts: the "river" still flows but at a rate so lacking in apparent motion as to be imperceptible through ordinary scientific investigation.

The speed of light, or the speed of the constant, defines the boundary between dimensions four and five, between the time dimension and the plane of light, where time as we know it slows down to a virtual standstill. Therefore speed, or rate of vibration, is the key. The faster you move, or the quicker you "vibrate," the slower time flows and, relatively speaking, the longer you live.

Now, if reaching the constant speed of light would gain us entry into the fifth dimension, one might suppose that the even higher sphere, the sixth, could be accessed in much the same way, but with one crucial difference—the "velocity" barrier, or the required rate of "vibrations," would no longer be the speed of the constant, but rather the square of it. This is no arbitrary choice of measure, of course. As we noted previously, it is one that is very subtly encoded in the Magic Square of Mercury, which in turn was associated with the Great Pyramid, known in ancient times as "The Lights," or "Lights-measures."

We have now made our way up to the penultimate sixth dimension and the timescale of the galaxy. So what happens here? Presumably time would still flow, albeit at a rate we can only describe as a virtual standstill squared.

From here we have one dimension to go, the ultimate seventh, the medium in which the whole exists. How fast would we have to move, or to resonate, in order to look out through the eyes of God into his seven-dimensional Garden of Eden? Once again, hermetic theory can provide us with a plausible answer: the cosmos is a highly ordered musical entity, and so the characteristic "vibrations" of each dimension must be harmoniously related to each other. Therefore if the plane of light is accessed through the velocity of light and the solid of reality is accessed through the square of the velocity of light, then the seventh dimension, the medium in which this "solid" exists, might reasonably be expected to unfold at the cube of this velocity. Here, "time" would truly stand still, and genuine nonlocality would be a living reality.

A BRIEF RETROSPECTIVE

Confused? To be perfectly honest, so am I. Frequently. But then we are trying to come to terms with the imponderable here, and left-brain logic alone can take us only so far in the quest for the ultimate reality. Eventually, it seems, we have somehow to experience this multidimensional reality for ourselves, and such experiences, as scientists are now aware, invariably involve a certain amount of intuitive insight. Unfortunately, this power, which might be considered the modern equivalent of the shamanistic vision, is a faculty that tends to appear only in sparse, random bursts. You cannot sit down and willfully intuit your way out of an intellectual maze; it just seems to happen. But these intuitive moments, these "macromutations" of the human mind, are essential to our evolutionary progress; they are the very life-blood of consciousness. Therefore not only should we pay heed to them, we should at all times be looking for possible ways to cultivate the soil in which they grow and thereby bring about an increase in their yield. The shaman uses mind-altering agents to induce such states; the mystic uses intense study, objective psychology, and rigorous discipline; the scientist simply relies on chance—hence the fragmentary nature of our accumulated knowledge.

As a consequence, the evolutionary process I am trying to envisage, with its plethora of scales and dimensions, very probably falls a long way short of the complete intuitive picture perceived by the originators of hermetic theory. I have personally had many vivid glimpses of this picture and have often felt overwhelmed at the sheer enormity of the implications of the central concept. At other times I have thought otherwise, that perhaps I might have lost the plot somewhere along the line and recklessly allowed myself to be carried along on the wings of my imagination. After all, who am I to pronounce on the theory of everything? What gives me the right to probe the disciplined mind of the modern evolutionist, the nuclear physicist, the theologian, even God himself? Am I not simply wasting my time dreaming up imaginary, incomprehensible worlds, when there are more practical things to do?

These and many other such thoughts have intermittently plagued me for years, but deep down I have maintained a conviction that the Hermetic Code is much bigger than me or my critics—or indeed all of us—and that it will forever continue to exert its influence on human consciousness irrespective of our individual prejudices and subjective experiences. I am therefore strongly inclined to come down squarely on the side of those who created this remarkable belief system, a genuine science, possibly the most highly evolved of Earth's inhabitants. These great visionaries were not only in tune with what Schwaller de Lubicz called "all the harmonies and energies of the universe," they were also profoundly altruistic and deeply concerned about the future development of mankind and of evolution per se. This is why they went to such great lengths to transmit their knowledge of the sacred laws of nature, because they knew that without it, without a clear understanding of the unity of everything, we should never be able to handle what lies ahead. And so the Hermetic Code is their legacy, a genuine seed of wisdom sown in distant times, whose "genes" have permeated the entire body of human consciousness. Not only does it explain exactly how and why the evolutionary process unfolds as it does, it also lets us as humans know how we can become a conscious part of it all, a truly remarkable gift to one's successors.

So the universe, as the ancient Greek initiates were saying two and a half thousand years ago, is hermetic throughout, very much alive, the direct organic result of some great cosmic act of procreation. We must therefore assume that, like all organic creatures, this great cosmic entity will eventually die. Scientific theory generally supports this view, i.e., that the universe was conceived, that it is growing at an unprecedented rate, that it will eventually reach its prime, and then start aging, eventually to dissipate all its energy throughout space and time in a long, slow burnout.

But of course death, according to hermetic theory, is not just the final "note" of one's personal evolutionary scale, it is also the first "note" of the greater scale above, the beginning, as it were, of a new, higher level of existence. The creators of the Hermetic Code, for example, died, like all of us must, but as we can all bear witness, their knowledge, their spirit, their higher selves, have lived on.

So let's suppose that the universe is destined to perish in what scientists see as a long, slow "heat-death." Will that really be the end of everything, as cosmologists predict? Or will the process of evolution extend yet further, as the universe's "higher self" transcends to greater things? Obviously on this point hermetic theory represents something of a departure from the scientific position, because it calls for an ongoing organic scenario, where the evolutionary processes above and below are seen as essentially the same.

These two opposing views might seem currently irreconcilable but, as we shall see in the following chapter, both theories, ancient and modern, conform to the same cosmic design. The main distinction between them lies in the fact that cosmological theory, like Darwinian theory, is concerned primarily with the physical body of the "organism," whereas the hermetic theory of transcendental evolution offers a much more holistic view, one that allows for the natural death of the physical body, but which in addition takes into account the wider, external influences created by the individual in life.

So now we find that the universe itself is also an "individual," and that it has definite and unique characteristics. As we noted earlier, galaxy distribution is not strictly homogeneous, which means that the "body" of the universe is lumpy and uneven like yours, the "seeds" of this design having been identified in slight variations in the background radiation left over from the big bang. As an individual, therefore, the universe, like you and me, may have a destiny as well as a fate. Fate is the inevitable lot of all organic creatures; it locks them into an irresistible life–death cycle that is beyond the individual's control. Destiny, however, is a potential, a future something that is developed and determined in an individual's lifetime and that is associated with the "higher self." In the case of the universe, we see that its fate is acknowledged by science, but that its destiny is left completely out of the picture. The next chapter is an attempt to rectify this imbalance by examining the established scientific viewpoint specifically in the light of hermetic theory. As we shall see, this exercise leads to some very interesting and rather startling conclusions.

fate vs. destiny!

11
The Fate of the Universe

We can begin here with the current scientific worldview, which holds that the universe is basically a chaotic, expanding mass of space, time, matter, and energy. The root of this assumption is the second law of thermodynamics, which says simply that heat always flows from hot to cold and never vice versa. This means that all the energy in the universe will continually disperse far and wide until it is distributed evenly everywhere. An involutionary process such as this, if it remains unchecked, will eventually lead to a long, slow, heat-death for the universe; all matter will become icy-cold and lifeless, its energy having been spread thinly throughout space, remaining only as a residual vibration, like a faint whisper in an otherwise silent void.

This rather depressing state of affairs, say cosmologists, should not concern us too much, because it will take many billions of years of gradually increasing entropy for such conditions to arise, and in any case the human race will certainly not be around to witness this ultimate state of chaos. This heat-death scenario implies, of course, that the universe is basically a closed physical system.

An alternative to the above prediction is the idea of the "big crunch." Proponents of this theory suggest that the collective force of gravity will eventually overcome all other forces. When this happens the entire cosmos will at first cease to expand; then it will begin to contract again under the cumulative force of gravity, increasing in temperature as galaxies converge, the whole thing collapsing inward toward a final "singularity"—the big crunch. This would mean the total annihilation of everything: space, time, matter, energy.

There has been further speculation that the resultant singularity could somehow trigger another almighty bang, so beginning a whole new cycle, a repeated expansion of the universe out to the limits permitted by the critical density of all its mass, only to contract again toward another mind-boggling crunch—and so on, ad infinitum.

Interestingly enough, this description is very similar to the ancient Hindu version of cosmological events, which sees the universe continually appearing and disappearing in a well-defined rhythmic cycle known as "a day and night of Brahma."

Physicists, however, doubt the possibility of endless cycles repeating without change, pointing out that there are serious physical problems with such a theory. We need not detail them here, but they apparently arise as an inescapable consequence of the inviolable second law, which would call for bigger and bigger cosmic cycles expanding with ever-increasing limits, until eventually future cycles would become so long that conditions within them would be indistinguishable from those prevailing in a big freeze.

Another interesting theory has been proposed by the science-fiction writer Wilbur Wright in his book *Time: Gateway to Immortality*. Wright begins by pointing out an interesting feature of the expanding universe, which is that the galaxies farthest away from us are receding at velocities close to one-tenth the speed of light. If these galaxies were eventually to reach the speed of light, he says, they might coalesce into enormous balls of matter and energy, ultimately contracting on themselves to become tiny, superdense neutron stars or black holes. Wright then goes on to propose a similar fate for the cosmos as a whole, suggesting that at the speed of light the tiny body resulting from the collapse of the entire universe might rupture the fabric of space-time and pass through into an adjoining continuum at high velocity and temperature. The end result would be another big bang, and so the beginning of another great cycle in the endless evolution of whatever it is that is evolving. Wright visualizes an infinite succession of universes and interpenetrating voids stretching from the unimaginably large to the infinitesimally small. As each continuum empties, a fresh singularity from some microcosmic region emerges to start a new cycle. As he says, nature abhors closed systems, so the sequence from small to large would be open-ended and potentially infinite.

There are, in fact, a number of alternative cosmogenic theories currently under consideration involving obscure phenomena like black and white holes, antigravity and inflationary processes, the "false" and "true" vacuum relationship, and so on. But clearly the most interesting of them all in respect to this present investigation is the proposition made by Lee Smolin, which is that there might even be a form of natural selection operating among universes, of which the evolution of life and consciousness may be a direct consequence. Smolin, of course, comes at this from the background of classical science. As we noted in chapter 9, the astrophysicist George Smoot has also opted for what looks set to become the "biological paradigm" by asserting that the structured physical characteristics of the present universe were already encoded within its "cosmic DNA" as early as a mere 300,000 years after the big bang. It is unlikely that these primordial "genes" simply materialized out of nothing, so we may reasonably assume that they were actually encoded within the original "cosmic egg" at the very beginning of time.

As we see, these ideas tie in perfectly with the hermetic description of events, which tells us that the universe above is a living creature, a zoon, and that the life and consciousness of sentient beings below or within it are faithful recapitulations of the original hermetic blueprint, part of an irresistible process that is vital to the sustained evolutionary development of the whole. Thus, while the physical body of this multidimensional creature may or may not be headed toward a final state of thermodynamic equilibrium, the greater universe, if it is alive and conscious in some mysterious way, could have emotional, psychological, and even spiritual sides to its existence. These are aspects we would define as being associated with the higher dimensions, in which the whole manifests as a six-dimensional phenomenon, an open system that, exactly like planetary biosystems below, is able to export entropy into its "environment"—the seventh dimension—and to simultaneously import the energy needed to sustain its ongoing development.

If this were so and the universal chain of existence proved to be open-ended and potentially infinite, then the second law would not be violated in any way: energy of some kind could still enter from outside the system. This external energy would not, of course, prevent the physical body of the universe from ultimately dying. Like your own, if it is basically organic, then eventually it must. But even if it were to die, through whatever means, one would still expect it to have the capacity to pass on its hereditary characteristics in some way. Possibly, therefore, the background radiation fluctuations described by George Smoot as "cosmic DNA" are the result of hereditary genes bearing the characteristics of some earlier parent universes.

THE "OTHER" UNIVERSE

As I have already implied, the transcendental "higher self" of the universe (that is its psychological and spiritual natures) would have to be connected in some way with the higher dimensions. The question is, How might such a connection be established? How could an expanding, chaotic mass of purely physical phenomena ever escape from the fourth dimension?

The distance to the edge of the universe, or to the outer wave of expanding galaxies, is not known, but scientists have calculated that these galaxies are moving away from us at around 10 percent the speed of light, as noted before. No one knows whether the outer galaxies will ever attain the speed of light, but if the universe has enough outward momentum to continue expanding at an accelerating rate forever, then it is not unreasonable to suppose that this could and very probably will happen. Attempts have been made to discover if there is enough gravitating matter in the universe to cause it to contract again, but there appears to be an unknown quantity of "dark" matter out there, so calculations have been necessarily speculative. However, Paul Davies has been moved to remark, in *The Last Three Minutes:* "Taken at face value, the galaxies seem to be flying apart so fast that they may indeed just 'escape' from the universe, or at least from one another, and 'never come down.'"[1]

Of course, if the galaxies did continue to accelerate unhindered they could eventually reach the speed of light itself, to the threshold of a quite different reality—the timeless, fifth dimension of existence. Wilbur Wright has suggested that at this point the whole galaxy might coalesce into an enormous ball of matter and energy, ultimately contracting upon itself to become a superdense neutron star or a black hole. However, this would be an unlikely end, one might think, for something so vibrant and radiant as a living cell in the body of the universe.

So, assuming this did occur, that the outer galaxies effectively "escaped" from the fourth dimension and reached the threshold of the plane of light, then theoretically they would be freed from the consequences of the second law of thermodynamics, which relies on the "arrow of time" to define any increase in entropy, or any waxing or waning of energy content. On the plane of light there would be no time mechanism with which to measure any kind of change. Out there it is always midday: nothing waxes, nothing wanes, everything just is. Presumably this is why the photon quantum, existing on the plane of light, is potentially everlasting; if unhindered by matter, it can maintain its vital spark and its maximum velocity for billions upon billions of years. The background microwave radiation permeating the whole universe, which scientists say is the residual vibration left over from the big bang, consists of photons—light quanta—and these quanta have been moving at the same maximum velocity from the moment they were first created. If these photons can last from fifteen to twenty billion years, they can reasonably be expected to last for another twenty billion years, and so on, in virtual perpetuity. Clearly there is no evidence of advancing chaos in such a dimension of existence, no increase in entropy as we would normally define it.

What we have here, of course, in this future scenario of galaxies escaping from the time dimension and "never coming down," is a graphic example of the universal process of transcendental evolution unfolding. These galactic helices, the cells of the universal body, would effectively enter into a higher scale of existence, a scale no less musical than the one below, but one that encompasses an infinitely greater reality.

So let us now trace the origin of the galactic cell back to the point at which the whole musical phenomenon first came into being. The big bang might be said to represent the very first note, Do, in the greater fundamental octave of universal evolution. Significantly, conditions of existence when this first note was sounded were such that there was no space, no time, and, therefore, absolutely no entropy. We can envisage this universal octave as having subsequently developed in all directions, from the first note, Do, of the big bang, up through various intermediary stages, perhaps into the "re" of early expansion and cooling, the "mi" of particle formation, the "fa" of the first star formations, the "so" of galaxy formation, the "la" of the formation of accompanying biosystems or habitable planets, and the "ti" of the appearance of organic life and of observers like you and me. These are merely hypothetical stages, but they all have one thing in common in that they manifest in the fourth dimension of time. In addition, they all encompass within their structures the three lower dimensions: solid, plane, line. Now, if we accept the hermetic interpretation of cosmic events, we can say that the time dimension, like everything else, is itself a fundamental octave, and that the ultimate note, Do, at the top of this scale of development would be sounded by all galaxies everywhere as and when they reached the light barrier—a kind of celestial version of the sonic boom.

In the case of the major musical scale, we know that the first note and the last are one and the same (Do), with a difference only in scale. By the same token we can say that the first and last notes of the universal time octave described above must also be in essence the same, again with a difference only in scale. Logically, therefore, one would expect the conditions prevailing at the moment of initial creation (no space, no time, no entropy) to prevail also inside galaxies entering the dimension above. And so they would, for at the speed of light, space contracts to nothing, time stands still, and everything moving at such a velocity—like the photon—is in a permanent state of thermodynamic equilibrium.

We have thus far followed the evolution of the universe from the big bang to the superluminal boom of "escaping" galaxies and so traced the development of one fundamental octave of universal resonance. According to musical theory, of course, the ultimate note, Do, at the top of this scale would not only be the last, it would also be the first note of the next scale above. Therefore, any galaxies developing up to this stage, transforming themselves, or a higher part of themselves, into five-dimensional entities would then have the potential to evolve up through the next ascending scale of universal resonance. So each galactic cell, upon reaching the threshold of the fifth dimension, would then continue to expand in some way, to develop further as its transcendental body approached nearer and nearer to speeds approaching the square of the speed of light, into a six-dimensional, solid form. Just like an individual cell in a growing planetary organism, the greater body of the galaxy would steadily become pregnant with six-dimensional cosmic "proteins." In the organic world, a normal self-replicating cell that has reached this condition of "optimum resonance" ultimately divides. Could a "pregnant" galaxy somehow do the same? If the universe is a closed system it probably could not, but if it is open and organic, living in a seven-dimensional arena, anything is possible. For evidence of such a process, which in our timescale might be an extremely rare occurrence, we should perhaps be looking for two galaxies coexisting side by side that are structurally mirror images of one another.

If a galaxy could "divide" in some way, the question as to how this might occur is even more perplexing. It has been suggested that at the center of all revolving galaxies is a black hole, a monstrous, superdense, supergravitational entity from whose clutches even light can't escape. A black hole would literally tear apart anything that came within its immediate sphere of influence—its "event horizon"—including, of course, stars and their planets. Certain astrophysicists have suggested that a black hole could act as a kind of nonlocal conduit through which anything passing might subsequently emerge "on the other side" into a totally different but coexistent space-time continuum.

It is hard even to imagine the extraordinary sequence of events that might transpire as the greater "body" of the galaxy transformed itself into a six-dimensional being. Once it had reached the first conceptual barrier—the speed of light—and transmuted some kind of resonance through to the fifth dimension, these transcendental vibrations would thereafter become a part of a whole, new, wider reality. Earlier we noted that a higher dimension unfolds every time a new direction is determined. In the case of the galaxy reaching the light barrier, this new direction would be something akin to a lateral, planelike development, spreading out at 90 degrees to the initial line of movement. The original line of movement would remain just that—a relative "line"—a cross-section of the greater five-dimensionality of the thing. Similarly the transition of the galaxy from five- to six-dimensional form would also proceed in an entirely new direction, so that its final condition would be as far removed from its five-dimensional manifestation as a solid is from a Euclidean plane.

So a galaxy that had successfully attained its final six-dimensional form would have effectively evolved up through the second fundamental octave of universal resonance, again from the first note, Do, sounded by crossing the light barrier, up to the next fundamental note, Do, the second barrier marked by the square of the speed of light. But even at this stage or on this scale, at the beginning of the third octave of universal development, the galaxy would still be in the process of evolving. We can accordingly depict this overall cosmic process of the galaxy's coming to fruition using our usual diagrammatic format:

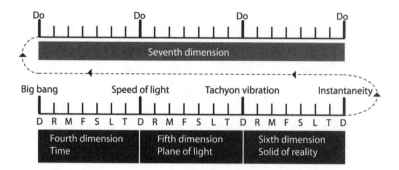

Obviously the mass of a galaxy could never accelerate beyond the speed of light. It would have to remain either suspended on, or marginally below, the plane of light. Any further superluminal motion, therefore, would have to involve nonmaterial resonances. These metaphysical vibrations would emanate from the material core, the galactic "chromosome," and would then proceed to develop in an entirely new direction, diffusing their energies, if not yet instantaneously, then at speeds far greater than that of light. As to the possible nature of these higher energies or forms of resonance, they might, as I have already suggested in earlier chapters, be composite structures formed from the conscious thoughts of sentient beings existing in the given galaxy. Admittedly we are in the realms of science fiction here, but then consciousness itself is a form of resonance, vibrant, energetic, alive, and its very existence represents a new direction in which the galaxy can continue to grow and develop.

So, if the galaxy itself is a chromosome, then, relative to the same scale of existence, sentient beings like ourselves would be the metaphysical equivalent of biochemical "triplet codons," created and subsequently ejected by the double helix above out into the cytoplasm (the planetary world in time), presumably so that we can ultimately dictate the synthesis of the higher, finer substances required for further evolution.

In the living cell, triplet codons act as templates for the manufacture of finer, more resonant substances—amino acids, the building blocks of life. Similarly, the human mind codes for the manufacture of metaphysical "amino acids"—ideas, concepts, theories, and so on. Now, in the cytoplasmic membrane of the cell, there are special enzymes that cause the newly developed amino-acid chains to fold up into the more complex protein macromolecules. Likewise, therefore, in the greater macrocosm there should be components out there in the cosmic "cytoplasm" of the sky whose function is to "fold up" our concepts and theories into immense, radiant, "protein" structures. As I suggested in chapter 7, these cosmic "enzymes" may be connected in some way with the sun and its planets (and perhaps their moons), whose varying magnetic influences pervade the whole solar system and whose orbital cycles are hermetically related to one another, endlessly beating out in time the relative values of the major scale. We note further, from Robert Temple's discovery that the Pythagorean Comma is expressed in the mass ratio of Sirius B and our sun, that this hermetic symmetry could also extend to the stars and even, perhaps, if the dictum of Thoth holds true at every level, to the galaxies themselves. Certainly starlight is hermetically structured, as indeed is all light—and it vibrates throughout the entire universe. So it would be no exaggeration to say that there is music literally everywhere, in the chromodynamic and atomic scales of matter, in DNA and the genetic code, in the double helix of the human brain and the Hermetic Code, in solar and galactic helices, in the octave of dimensions—even in the "mind" of the universe itself.

12

Inner Octaves

By this stage readers will appreciate that the universe may have many more facets than science currently allows. No longer do we see it as simply a four-dimensional phenomenon involving according to thermodynamic principles; it has now become a vibrant, essentially six-dimensional entity, possibly teeming with innumerable kinds of lesser six-dimensional life forms. These various life forms, as we have noted, occupy various scales of existence on the evolutionary ladder, beginning at the level of DNA and culminating at the scale of the galactic helix, all of them coexisting within a framework of seven dimensions.

Now, this ascending "ladder" is not simply a progressive chain of separate rungs placed one on top of another. If the whole universe is a living entity, this means that it is a fully synchronized body, the vibrations of all scales interpenetrating and reinforcing one another strictly according to the dictates of the grand design. We might best view this evolutionary phenomenon as a series of seven pulsating spheres of vibrations, each being contained within the one above it, all of them sharing the same central point.

For example, linear DNA contains within it the whole atomic scale, an infinity of "points," or an endlessly variable sequence of nitrogenous base pairs, each consisting of a few fundamental atoms. But it also contains within it the seeds, or genes, of the greater scale above; it is the blueprint, the recipe, for the creation of the entire organism. In the same way the organism of the human being contains within it the whole DNA scale, an "infinity" of biomolecules, and also, one assumes, the seeds or genes of the greater solar body above. We can imagine the same process repeating itself up through the galactic scale, to the ultimate, Absolute scale.

The hierarchy of dimensions is also integrated in the same manner. A one-dimensional entity—a line—contains within it an infinite number of zero-dimensional points and is a cross-section, a blueprint, of a greater plane; a plane is comprised of an infinite number of one-dimensional lines and is a cross-section of a greater solid; and a solid, in a similar fashion, contains within it an infinite number of two-dimensional planes and is a cross-section of a greater four-dimensional entity existing along the line of time. Exactly the same pattern would repeat itself in the metaphysical scales above, where the four-dimensional line of time encompasses all three-dimensional possibilities, the five-dimensional plane of light all four-dimensional possibilities, and the six-dimensional "solid" form of the ultimate reality embraces everything: points, lines, planes, solids, time, eternity.

Such a view expresses above all the holistic nature of the universe, on which we shall be concentrating in the following two chapters. We are now familiar with the idea of the complete interconnectedness of everything, a principle that mystics and yogis have intuitively understood for thousands of years and which scientists of the twentieth century latterly discovered through the so-called nonlocal quantum correlations existing between widely separated particles. But is there a way in which this somewhat tenuous and abstract reality can be better understood? That is, if the entire universe is a nonlocal arena of interpenetrating and mutually interacting vibrations, how might such an all-encompassing process work? For example, how can vibrations or wave/particles in one part of the universe be simultaneously "in tune" with vibrations light years away? Or, alternatively, how could the conscious mind of the mystic or the shaman or the LSD tripper connect with a nonlocal reality?

In chapter 8, we noted that musical theory itself provided at least part of an explanation for simultaneity, whereby the ultimate note of any given harmonious scale can at one and the same time exist in other scales, above and below. But can we determine what kind of mechanism allows these vastly different scales to be so intimately linked?

As it happens, we can. And, not surprisingly perhaps, we need look no further than the theory of transcendental evolution, the "theory of everything," for a major clue. This is the sacred number 64, the number of infinite harmony, the key, as it were, to infinity. Primarily associated with the Great Pyramid—a monument dedicated to light, or "lights-measures"—the number 64 tells us that an octave of light is further subdivisible into eight inner octaves.

Just for the record, this concept of inner octaves—an outline of which follows in a moment—did not come to me directly as a result of my preoccupation with the Hermetic Code. In fact, I first came across it several years before I fully realized the Code's significance. My source at that time was Ouspensky's *In Search of the Miraculous*, the record of lectures given by George Gurdjieff in Moscow and St. Petersburg at the turn of the twentieth century. As I said in the introduction, Gurdjieff always claimed that his system of knowledge was drawn from teachings reaching back into the remotest antiquity, but even after reading everything written by or about him it was some time before I made the connection and realized that the principle of inner octaves is in fact very neatly embodied in the Hermetic Code, the oldest recorded teaching on Earth.

Gurdjieff tells us that all matter vibrates, resonates within, in the form of octaves. Normally, when we speak of matter, we are referring to phenomena of substance, things we can touch, see, or measure through some form of scientific method. According to Gurdjieff, however, the property of materiality spans the entire universal spectrum. "Everything in this universe can be weighed and measured," he said, "The Absolute is as material, as weighable and measurable as the moon, or as man."[1] The higher orders of materiality, however, are much too rarefied to be regarded as matter from the point of view of chemistry or physics; matter on a higher plane is not material at all for the lower planes, but it permeates them nonetheless.

In his lectures, Gurdjieff often referred to a cosmological model known as the "ray of creation," which, he said, belonged to ancient knowledge. Basically, it was an elementary plan of the universe, beginning with the highest "world order" and ending with the lowest, so:

1. Absolute
2. All worlds
3. All suns
4. Sun
5. All planets
6. Earth
7. Moon

As we see, the ray of creation, like the hierarchy of dimensions already discussed, represents seven planes in the universe, seven worlds, one within another. (Pythagoras, incidentally, expressed this same view through his geometrical symbol known as the Lambda, comprising seven concentric circles.)

Gurdjieff then described this descending octave, or order of worlds, in terms of the cumulative effect of the law of three forces at each successive level. In the world of the Absolute, the three forces, being harmoniously related in the fullest sense, constitute one whole. The Absolute world is therefore designated by the number one.

In a world of the second order (all worlds), the three forces are already divided. Such a world would be designated by the number 3. These three divided forces, meeting together in each of these worlds, create new worlds of the third order (all suns), each of which manifests three new forces of its own, so that the number of forces operating within them will be six. In these worlds are generated worlds of the fourth order (the sun), in which there operate three forces of the second-order world, six forces of the third-order world, and three of their own, making twelve forces altogether. The process continues, giving twenty-four forces in worlds of the fifth order (all planets), forty-eight in worlds of the sixth order, of which the Earth is a part, and ninety-six in the seventh (moon). It follows, therefore, that the number of forces in each order of worlds, one, three, six, twelve, and so on, indicates the number of laws controlling it. So the fewer laws there are in a given world order, the nearer it is to the will of the Absolute; the more laws there are in a given world, the greater the mechanicalness, the further it is from the will of the Absolute. We live in a world subject to forty-eight orders of laws, that is to say, very far from the will of the Absolute and in a very remote and dark corner of the universe.[2]

Following on from this descending pattern of accumulating laws and forces, Gurdjieff then explains how the materiality of each world order differs accordingly, becoming ever denser as it involves from the Absolute to the moon. All matter, he says, including that of the world of the Absolute, is composed of "primordial atoms." Obviously these "atoms" should not be confused with those described in ordinary physics; rather, they are certain small particles that are indivisible only on the given plane. Only on world 1, the world of the Absolute, are these particles truly indivisible. The "atoms" of world 3 consist of three atoms of the Absolute world and so would be three times bigger and heavier. Again, the "atoms" of world 6 each consist of six atoms of the Absolute—and so on, according to the laws and forces described above, with twelve primordial "particles" constituting an "atom" of world 12 and a corresponding increase in density as we pass further down through worlds 24, 48, and 96. We thus have seven different orders of "materiality" in the universe. Our ordinary concept of one order, said Gurdjieff, just about embraces the materiality of worlds 96 and 48. The substance of world 24, he said, is almost too metaphysical to be identified through ordinary scientific method; and the even more rarefied substances of worlds 12, 6, 3, and 1, have, to all intents and purposes, no identifiable material characteristics.

It is interesting to note here that in the early 1900s when Gurdjieff was giving his lectures, the conventional atom was still the smallest "particle" of matter known to science. But while Gurdjieff was speaking of these still finer substances permeating the material world, Ernest Rutherford was discovering the nucleus of the atom, Einstein was attempting to show that photons were particles, and Max Planck was in the process of formulating his idea that electromagnetic radiation was emitted by energetic sources in discrete, symmetrical packages called quanta. Later discoveries, such as the existence of neutrons and electrons within the atom, and of quarks literally everywhere, all served to reinforce Gurdjieff's idea that there are finer substances and vibrations permeating the coarser ones. Nowadays, of course, even individual quanta, the tiniest "particles" known to science, are also described as ephemeral, wavelike entities, suggesting the existence of a finer, more rarefied manifestation of "materiality" than even the most minuscule, pointlike quantum.

Clearly the claim that all matter everywhere is actually composed of the fundamental and indivisible particles of the Absolute world has no arguable scientific basis. As Gurdjieff himself said, the substances of the higher worlds have no recognizable or measurable material characteristics. On the possible nature of such materiality, however, we can speculate at least as far as world order 3 (dimension six in the hierarchy), the world of "all worlds."

Beginning with the lowest world order in the ray of creation, the moon, or world 96 (an order of materiality that would also incorporate the interior of planet Earth), we can say that the matter of this world would probably consist for the most part of the heavy transition metals and all the superdense radioactive elements.

The next world order in the ascending scale, represented in the ray of creation by our own planet and its atmosphere (world order 48), would incorporate matter comprising atoms of the lighter chemical elements, ending with hydrogen, the least dense of them all. Accordingly the matter of the world of "all planets," world 24, might consist primarily of subatomic particles. Beneath the "particle," be it photon, electron, or whatever, lies the even more rarefied wave-mode vibration. Let's say that this wave aspect of subatomic quanta represents the nature of the materiality of world order 12, the world of our sun.

We have now come to the outer limits of scientific knowledge. On the reality beyond the wave we can only speculate. Hermetic theory tells us that even finer vibrations exist within these waves. Possibly, therefore, the next order of materiality, that of world order 6, "all suns," is consciousness itself, the "substance" from which, as I suggested in an earlier chapter, all solar helices are constructed. The materiality of the next world order—the scale of the galactic helix—might be defined as a form of superconsciousness (ordinary consciousness squared, as it were), a substance that, if it exists, must be so rarefied that it must forever remain hypothetical. Finally, the materiality of the primordial "atoms" of the Absolute world order, as we might expect, defies all expression.

We now come to the concept of inner octaves. According to Gurdjieff, each note of any given octave can be regarded as a complete octave on another plane. Similarly, each note of these inner octaves is also a complete octave in another scale—and so on, but not ad infinitum, because there is a definite limit to the development of inner octaves (just as there is a definite limit to the hierarchy of dimensions and the ray of creation). These inner vibrations, said Gurdjieff, proceed simultaneously in media of different densities, continually interpenetrating and interacting with one another. In a substance or medium consisting of, for example, the superdense atoms of world order 96, each of which is a composite of 96 primordial "particles," the vibrations or oscillations active within this medium are divisible into octaves, which are in turn divisible into notes. The medium of world order 96, like a solid piece of wood saturated with water, is also saturated with the substance of world order 48. Now, the vibrations subsisting in the matter of world order 48 stand in a definite relation to the vibrations in the substance of world order 96; each "note" of the vibrations of world

96 contains a whole octave of vibrations in the medium of world 48. These inner octaves, said Gurdjieff, proceed inward to the very heart of all matter. The substance of world order 48 is in turn saturated with the substance of world order 24, so that each "note" in the vibrations of world 48 again contains a whole octave of the vibrations of world 24—and so on through to the final phase, where the substance of world order 3 is permeated with the substance of world order 1, with each note in the vibrations of world 3 containing a whole octave of the vibrations of the world of the Absolute.

As I mentioned before, Gurdjieff always claimed that the original teachings from which his ideas were drawn—including the above description of inner octaves—dated back to very remote times. How far back this teaching actually does go is currently the subject of much heated debate among alternative theorists and orthodox historians, but it was very much alive in Old Kingdom Egypt, as we know from the previously discussed Magic Square of Hermes and its associate number, 2,080, the sum of all the factors from 1 to 64. Obviously 64 is the key. The Greeks, as we know, associated the Magic Square with the Great Pyramid, "The Lights." And light, of course, is an octave of resonance, composed of eight fundamental "notes." According to Gurdjieff, each of these fundamental notes in an octave of ordinary light would contain a whole octave of notes from the scale or world above. As we see, this very principle is precisely encoded in the Magic Square.

Gurdjieff claimed that "objective music" (by which he meant the kind played by such as Joshua and the builders of the Egyptian and Orphic schools, which allegedly could move mountains of stone) was all based on these inner octaves. Ordinary music, he said, cannot be used to reconstitute matter, destroy, or build up great walls of stone, but objective music can.

The music being referred to here is, I believe, fundamentally psychological music, the music of the mind, the music described by the Hermetic Code and the I Ching, by the Greek and Egyptian mysteries, and, indeed, by the established principles of all major religious disciplines. In Egypt, this "religion," the making of "celestial music," was known as "writing," the sacred art invented by Hermes/Thoth, the art of striking harmonious metaphysical "notes," or thought patterns, up into the stellar scale of existence, into the "heavenly" world inhabited by the gods. We must assume here that this does not mean "writing" in the ordinary sense.

So let us just imagine for a moment that the mind were conscious to the degree that it could generate higher vibrations—inner octaves—that were in tune with solar helices, world order 12 in Gurdjieff's ray of creation.

As I have suggested, this level of materiality would be as fine and as penetrating as the ghostlike wave mode of subatomic quanta, reaching, as it were, beyond the particle itself into the very heart of the electron. It is not too difficult to imagine some kind of process whereby such vibrations, if they could be concentrated or focused to a sufficient degree of intensity, could indeed have dramatic psychic and physical consequences. Theoretically such rarefied "substances" could actually enter into objects—even blocks of the hardest stone—and affect them from within.

No doubt most orthodox scholars will regard such a notion as entirely fanciful, but not, I would hope, all of them. Times are changing, and scientists are today having to rely as much on intuition and instinct as they are on logical cognition in their attempts to come to terms with the baffling nonlocal nature of the multidimensional universe. We might optimistically view this scientific venture beyond the empirical world out into the metaphysical realm of concepts, thought patterns, and vibrations, as evidence of evolution of the transcendental kind, the beginning of mankind's next momentous journey—to the stars. If this is so, then the rationalist, whether knowingly or unknowingly, may now be contributing actively toward this ultimate flowering of human consciousness.

Take the ideas of David Bohm, for example, the "orthodox" scientist mentioned in chapter 4, whose investigations into plasmas led him to conclude that the electron is a "mindlike" entity. We may recall that he felt instinctively that the "plasmon"—the electron sea—was alive, with billions of individual electrons simultaneously engaging in a mass, instantaneously coordinated action. This implies that electrons are somehow able under certain conditions to "connect" with every other electron, and Bohm recognized that the nonlocal nature of interactive quanta could account for this kind of synchronized activity.

Impressed by the evidence for nonlocality, Bohm went on to develop what at first appears to be a revolutionary new view of the universe. He suggested that the whole of reality was like a living hologram, a "holomovement," and that what we see through ordinary methods of investigation is something like a frozen holographic image, behind which lies a much deeper and more meaningful level of reality. Now this idea may be new to science, but it is revolutionary only in the sense that it has turned full circle: it has been held before. In fact, this "holographic principle," as we shall see, is basically an updated scientific description of the mechanism of inner octaves and of the principles of musical theory.

13

The Holographic Principle

Most readers will know that a hologram is a three-dimensional image sculpted from a concentrated beam of light. The thing that makes this possible is the wave-mechanics phenomenon known as interference. We came across this in chapter 4, where we discussed the Thomas Young experiment, in which waves of light passing through twin slits in a partition overlapped and reinforced one another, producing an interference pattern on a dark backdrop.

In the same way, a laser, which is a very pure form of light, can be used to create extremely well-defined interference patterns. The hologram is produced by splitting a single, concentrated beam of light into two. One beam is then reflected off the object being photographed and the second beam is directed at an angle toward the reflected light of the first. The interference pattern created by the two beams is then recorded on film. The image on the film actually bears no resemblance to the hologram it projects, however. Only when another beam of light is shone through it does the hologram appear.

Impressive as these images can be, the most interesting aspect of holography concerns the film itself, which possesses rather unusual properties. Let's say we have a piece of holographic film on which a certain image has been recorded. It can be any image you like. If you were to cut this piece of film in half and shine a laser light through any one of the two pieces, you would find that each separate piece would still contain the whole image. Even if you cut each half into quarters, eighths, and so on, each diminishing piece, when illuminated by laser, will still project a complete image of the object in focus, albeit becoming progressively less distinct as the pieces get smaller. So every small segment of a piece of holographic film contains all the information contained in the uncut whole. This is the holographic principle.

In Bohm's view, the physical world we see all around us is just like a holographic image, basically an illusion, a kind of external tapestry of subjective impressions composed of waves and interference patterns. Beneath these tangible physical forms, he suggested, lies a deeper, "implicate" order of reality, in which everything exists in what he called its "enfolded" form. Therefore what we see as physical phenomena are simply the explicate or unfolded projections from this deeper, implicate, enfolded order.

In fact, this view of two fundamental orders of existence—i.e., of the "image" and the "film," or the explicate and the implicate—has exact parallels in numerous esoteric traditions. For example, Buddhists call the material world the sphere of the nonvoid. This is the normal world of sense-objects, the explicate, unfolded dimension—the dimension in which the "holographic" image manifests. The real world, existing beyond the nonvoid, is the void, the dimension in which the "film" itself exists, the implicate, enfolded realm, the progenitor of every thing, every "image" in the visible universe.

Michael Talbot, author of *The Holographic Universe,* which provides a general summary of Bohm's ideas, quotes the Tibetan scholar John Blofeld speaking on the nature of the two domains. Blofeld's worldview, as we see, is strikingly similar to Gurdjieff's: "In a universe thus composed, everything interpenetrates, and is interpenetrated by, everything else; as with the void, so with the nonvoid—the part is the whole."[1]

A similar Hindu version of cosmic events mentioned previously describes the motion of the universe as cyclical—an endlessly unfolding and enfolding process, with each cycle lasting "a day and a night of Brahma." It is perhaps worth noting here that the Hindu creation myth says that twenty-four "Brahman hours" are equivalent to 4,320,000,000 of our years, while four, three, and two, followed by seven zeros is a perfect description of the evolutionary development of the Pythagorean Tetrad, which Pythagoras himself, remember, referred to as "the model of the gods."

So the visible universe is created by Brahma. Brahma is one of the three major gods of the Hindu "trimurti," which itself, like all religious "trinities," is primarily an expression of the first law of nature, the law of three forces. Brahma therefore represents the first, active force in the process of triple creation, a force that originates in the implicate realm. From Brahma, everything enfolded subsequently unfolds, like a holographic image, into the explicate dimension. At the end of the "day" the world is "destroyed," or absorbed, by the god Shiva, the passive or negative force, which, in Bohm's terms, means that the unfolded again enfolds, from the explicate back into the implicate dimension, after which it completely disappears into the body of the third god, Vishnu, the omnipresent neutral force, the great cosmic mediator. Vishnu then sleeps for a "night" in the sphere of "non-existence," which Hindus describe as the dimension of endless time, and then gives birth to Brahma again. A new "day" unfolds, and the process endlessly repeats itself. Vishnu's role in this cosmic episode thus implies that there is an even deeper level of reality beyond the implicate realm, what Bohm himself referred to as a kind of "ultra-implicate" reality.

We have, of course, already visited this rather special place; it is number 6 in the hierarchy of dimensions, the dimension existing beyond the fifth, the plane of light—what I have referred to as the "solid" form of the ultimate reality, the sphere of true "nonlocality." In Gurdjieff's "ray of creation," this would be world order 3 in the descending scale.

Like the Hindu description of the universal process, Bohm's takes into account the fact that things are never static. He saw the whole phenomenon as in motion, hence his use of the term *holomovement*, which was meant to include not only the evolving universe but also, crucially, the consciousness of the observer. So when we see things through Bohm's eyes, we are not merely looking at the ever-changing hologram, we are an integral part of it.

Let us now consider the holographic principle itself, the fact that every small segment of a holographic film contains all the information in the uncut whole. Bohm's hypothesis implies that we as individuals, each an intrinsic part of the entire holomovement, must also, just like a segment of a holographic film, contain within us a complete "picture" of the greater reality.

This, of course, is precisely the reality Gurdjieff described decades before in his discourses on inner octaves. He said that if one understands the laws governing the creation of inner octaves, it is possible, from observations made in just one scale, to obtain the measurements of any other scale, because they are all in a definite relationship to one another. Therefore there is no need to study the sun, for example, in order to discover the nature of the "matter" of the solar world, because this same order of materiality exists in ourselves. In the same way we have in us the "matter" of all scales, for man is in the full sense of the term, a "miniature universe"; in him are all the matters of which the universe consists; the same forces, the same laws that govern the life of the universe, operate in him; therefore in studying man we can study the whole world, just as in studying the world we can study man.[2]

Moses expressed the same idea more succinctly: "for in the image of God made he man."[3] And again, the holographic principle—the whole existing in every part—can also be understood in terms of the basic rules of ordinary musical theory. Each tone, semitone, quarter tone, and the like of a major scale contains within it all the information necessary to recreate the entire scale. That is, to determine the frequencies of all the sounds comprising a harmonic scale or octave, it is sufficient to fix the frequency of just one of them.

So the holographic principle, like so many other scientific discoveries discussed in this book, is not an entirely new concept. It is simply a variation on one aspect of a very old theme, the essence of which is encoded in the Hermetic Code and in the number 64. Sixty-four is the key to the inner octave, the mechanism through or by which, said Gurdjieff, all scales of existence are connected and proportionately interrelated.

The holographic model of the universe raises intriguing questions in the spheres of psychology and parapsychology.

First, we now have neurophysiological evidence to suggest that the brain itself may have holographic properties, in the sense that such functions as vision and memory are distributed evenly throughout its structure. If true, this would be significant in serving to reinforce Bohm's idea that we are all part of an immense hologram in motion, a holomovement, and also Gurdjieff's assertion that we are exact replicas of the greater whole. Obviously, if the greater whole is by nature holographic, a holographic brain would be precisely what, in Gurdjieff's worldview, is required.

The holographic model also has possible parapsychological implications. For example, in an attempt to explain how psychokinesis might work—an uncharacteristically bold move for a mainstream scientist—Bohm cited the "ultra-implicate" dimension as the most likely source of such forces. This would be dimension six in the ascending hierarchy. The implicate dimension of Bohm's vision of reality would therefore correspond to the fifth, the nonlocal plane of light, with the explicate sphere—dimension four—corresponding with the line of time, expressed hermetically in Hindu myth as a period of 4,320,000,000 years.

So, like Gurdjieff, the Greeks, and the Egyptians, Bohm believed that the human mind could in fact access this higher dimension—the ultra-implicate, six-dimensional abode of Vishnu—and through it directly influence the physical world.

We are now back to the idea discussed in chapter 3, in which we considered the possibility that the ancient builders of the first civilizations might have moved their giant blocks of stone using some kind of psychic assistance.

Mind over matter? Admittedly the proposition sounds fantastic, but then so did the prospect of men on the moon little over a decade before it became a reality. Indications are that the next stop could be Mars or possibly one of the moons of Jupiter—a much greater feat than a hop to the moon, but few people today doubt that this kind of enterprise is within our capabilities. Of course the developmental leap from my mind or yours to one capable of defying the known laws of physics would be massive indeed. And yet the myths, the yogi masters, Gurdjieff, and latterly David Bohm all speak of such powers being accessible to the human being. Gurdjieff and Schwaller de Lubicz, as we noted, both believed that such powers were common currency in the ancient world and that somewhere along the trail of time our ancestors somehow lost the understanding of how to use these powers. Whether such a "Golden Age" actually existed or not, in light of the archaeological evidence we have discussed, where blocks of stone weighing from two hundred to twelve hundred tons have been carved, transported for miles, and then perfectly placed and oriented to form massive symbolic structures, I personally believe that ruling out some kind of hitherto unrecognized psychic or psychological factor in the lives and works of these ancient peoples would be injudicious.

As we have seen, the idea of mind—or "music"—over matter is as old as civilization itself, having originated with the men-gods of mythology. We noted Graham Hancock's observation of certain Native American myths that tell us that music—whistling, the playing of trumpets—caused heavy blocks of stone to float through the air like feathers in a breeze. Exactly the same kind of stories appeared in ancient Greece, recounting the exploits of such as Orpheus, son of the god Apollo and the muse Calliope, whose playing of the lyre "enchanted the trees and rocks and tamed wild beasts."[4]

Another hero was the "builder god" Amphion, son of Zeus and

King of Thebes, also a musician, who single-handedly built the walls of his great city. So clearly, as Gurdjieff always maintained, music of some sort is the key to the techniques used by these people. He further stated that such music involved the use of inner octaves, an all-pervading symmetry of composition that, he said, permeates everything, both man and the universe. Possibly there was real music involved in the procedure, or at least sound vibrations, as in case of the Tibetan demonstration described in chapter 3, which allegedly involved the use of numerous drums and trumpets. But in addition there may have been some form of psychological accompaniment, and it is here, one suspects, that Gurdjieff's inner octaves would come into play. Inner octaves or "higher vibrations," according to Gurdjieff's view, can be accessed only by a fully conscious mind; they correspond to an extremely high degree of "psychological resonance," a unique condition of existence, apparently attainable by ancient man, now surviving only as a potential faculty in the form of a lingering memory enshrouded in myth.

Scientists in general have a natural tendency to react negatively when they hear talk of "vibrations," particularly when what is being alluded to is an immeasurable commodity. And yet, consciousness itself, whatever else it may be, can reasonably be considered as a form of resonance. We can't measure these invisible forms of resonance directly, but we are, nevertheless, acutely aware of their puzzling existence in a world as yet barely half understood. Certain aspects of the rational side of consciousness, such as IQ, intellectual argument, ideas, theories, and so on, can be roughly appraised, but the nature of other aspects of the mind, such as its power to intuit, remain tantalizingly beyond our understanding. So, if what we might call rational consciousness can be conceived of as a form of energy manifesting in varying "degrees of resonance," then a faculty such as intuition, which clearly transcends all logical thought processes, must function with even finer "vibrations," that is with higher degrees of psychological or psychic resonance.

The same might be said of memory, a faculty that enables us to pick out a familiar face in a crowd even though thirty years might have ravaged it since last we saw it. It is invariably a very different face, and yet it is the same one, and the brain can somehow recognize this, it can decode highly complex, personal information that no computer could ever handle; it can "filter away" the lines and the scars, the changing hues, and all the rest of the camouflage of the years and simultaneously "see" the original face with pristine clarity. This faculty is so familiar to us that we barely give it a second thought, and yet it really is quite remarkable, way beyond the reach of modern technology. But it is also much more mercurial in nature than ordinary thought processes, such as the ponderous form of logical cognition required to write a page of a book like this.

Unlike intuition, memory can, in fact, be experimentally observed, and important new neurophysiological research now suggests that the brain cells or whatever else is responsible for this extraordinary faculty are not housed in any particular region of the brain, but are distributed evenly throughout its structure. This suggests that memory at least, one of the primary functions of the brain, is a manifestation of the holographic principle, the principle of inner octaves, where the whole exists in every part.

14
Quantum Psychology
The "Nonlocal" Brain

The idea that the brain functions on some kind of internal holographic principle was first suggested by a Stanford University neurophysiologist named Karl Pribram. Pribram was initially concerned with memory, how it works, and how the brain manages to store it. At the time he began his research, well over fifty years ago, it was thought that memories were localized inside the brain in the form of imprints known as engrams, chemical codices thought to be housed within specialized brain cells or biomolecules.

Up to the present, engrams remain only hypothetical entities: none have been identified or located, and Pribram began to doubt their very existence during his 1940s work with the neurophysiologist Karl Lashley at the Yerkes Laboratory of Biology in Florida. At that time Lashley was experimenting with rats trained to perform various tasks, like finding their way through a maze. He attempted to cut out the region of the rats' brains in which the memory of their learned skills was thought to be encoded, but he found that no matter what section of the brain he surgically removed, the rats still retained their memories. Even if their motor functions were chronically affected, they still managed to negotiate the mazes successfully and find their way to the larder.

From these findings, Pribram concluded that memories were not localized in specific areas, but were somehow distributed throughout the entire brain. He puzzled over this for many years, wondering how the brain could store memories intact throughout its whole structure. So the construction of the first hologram had a great impact on him, because it seemed that the process of holography, which results in an image of the whole existing in every part of the film, provided a plausible explanation of the nonlocal nature of memory.

Further experimental evidence in support of Pribram's ideas resulted from the work of Paul Pietsch, a biologist researching at the University of Indiana. Pietsch's work involved somewhat macabre experiments, primarily on salamanders. He found that he could extract a salamander's brain without killing it, leaving the creature in a torpid state; when he replaced the brain, the salamander's physical functioning quickly returned to normal. In a subsequent series of several hundred operations, he systematically chopped and removed different parts of the hapless creatures' brains, shuffling the right and left hemispheres, turning them upside down, back to front, even mincing them. But when he replaced what was left, he was astonished to find that their behavior always returned to near normal.

Skeptical at first of Pribram's claim that memories are not focused on specific brain sites, Pietsch ultimately concluded that this must be so, otherwise a minced brain would surely result in a correspondingly uncoordinated series of equally "minced" motor functions. The fact that this clearly was not the case led Pietsch to the opinion that Pribram was right after all: that the holographic model currently provides the best explanation for such an otherwise inexplicable property of the brain.

Pribram found further evidence to support this theory in another of Karl Lashley's discoveries, made during his experiments with rats, which indicated that vision might also be holographic. Lashley found that even after major surgical plundering, the nerve complexes controlling vision could still function normally. As much as 90 percent of the visual cortex could be extracted, yet the rats persistently retained their visual powers. It was subsequently discovered that the same was the case with a cat's optic nerve, 98 percent of which could be severed without seriously affecting its vision.

Previously it had been assumed that there was an exact correspondence between the images seen by the eye and the resultant pattern of electrical activity taking place in the visual cortex: that is, if you looked at a certain physical shape, the same image would be projected onto the surface of the cortex, like a photographic imprint. To find out if this was the case, Pribram conducted a series of experiments to locate and measure the electrochemical reactions in the brains of monkeys as they carried out a number of visually centered activities. He could find no identifiable pattern in the distribution of electrical activity, so it was evident that the visual cortex was not operating on a one-to-one basis with the image it recorded. This fact, together with the strange ability to continue functioning relatively normally even after drastic surgical excision, led Pribram to conclude that vision, like memory, is distributed evenly throughout the brain, which processes visual information using some kind of internal holographic principle. This would explain why even a small segment of the visual cortex is still able to construct everything the eye sees. As Michael Talbot points out in *The Holographic Universe,* the interference patterns on a piece of holographic film bear no discernible relationship to the images encoded on it. If the visual cortex were similarly functioning holographically, this could account for the fact that there is no one-to-one correspondence between the image seen and the pattern of electrical impulses activated on the surface of the brain.

Pribram believes that the brain could be using wave patterns to create these internal "holograms." Active brain cells (neurons) radiate electrical impulses from the multiple ends of their branchlike antennae, which expand outward like ripples in a pond. Electricity is in essence a wavelike phenomenon; therefore, as the impulses spread throughout the brain, they must be creating an overall web of interpenetrating waves and interference patterns. In Pribram's view, it is this wavelike interconnectedness that gives the brain its holographic properties.

Now, according to Bohm, the observer and the observed—the holographic mind and the holographic universe—should in no way be considered as separate entities, but more as interacting coordinates of the self-same "holomovement." This in turn implies that some kind of connecting principle exists between the two, and the terms that are now most frequently used to account for this possible function are: vibrations, resonance, waves, and interference patterns—all words, in fact, that are used to describe events in the nonlocal world of the quantum physicist. It is for this reason, and not because it is fashionable, that I use the term *quantum* to describe the kind of psychology that might be involved in connecting with the greater whole. And as we have noted, this nonlocal reality is strikingly similar to the world described by Gurdjieff, a greater sphere in which everything is seen as being interconnected through the "holographic" mechanism of inner octaves. This is also, as I suggested in earlier chapters, the "eternal" world—the Duat—of the ancient Egyptians, who regarded the phenomenon of light, the prime mover in the nonlocal, quantum world, as sacred, as an octave of resonance, each note of which is composed within as an octave, giving sixty-four interpenetrating "notes." Clearly, therefore, this notion of interpenetrating vibrations, intrinsic to the world of the quantum physicist, is one of the oldest testaments on Earth.

Bohm was obviously not what we might call a run-of-the-mill physicist. To begin with, practically alone among his peers, he was quite prepared to tackle the prickly subject of psychokinesis, "mind over matter," a proposition that has been a complete anathema to most scientists ever since Newton discovered what were long considered to be inviolable laws, the fundamental physical laws of motion and gravitation. Basically Bohm believed that psychokinesis might result directly from the essential common feature of both consciousness and the fundamental wave/particles of matter: an underlying "awareness" of certain information relating to the world at large. Like you, electrons and photons have the ability to respond to meaning, or to make positive use of external data. Bohm likened the process in the microworld to that of a ship on automatic pilot, where the radarlike wave function of the electron, for example, provides the particle aspect—the "ship"—with information about its environment. The implication is that the frequencies at which the "radar" works can be tuned into by the mind; that, in effect, the mental processes of one or more people could possibly be focused on frequencies of resonance that are in concert with the generative vibrations controlling material systems. By their very nature, such processes would involve forces other than those currently known to physics. They would arise as a result of what Michael Talbot calls a nonlocal resonance of meanings, a kind of interdimensional alchemical dialogue between mind and matter—something like the nonlocal alchemy taking place between correlated photons, or electrons in plasmas, but possibly involving resonances of a much higher or finer frequency. Therefore, in order to accommodate psychokinesis and perhaps other inexplicable phenomena such as telepathy, precognition, and so on, "ordinary" nonlocality must be superseded by what Talbot calls a "super non-locality"[1]—which in Hindu terms might be described as the unknowable process operating in the hidden world of the Great Mediator Vishnu, described as the sphere of "endless time." This would be Bohm's "ultra-implicate" sphere, our sixth dimension.

We noted previously that Gurdjieff regarded the processes involved in psychokinesis in much the same way, that is, as the result of a mutually interacting resonance between mind and matter. But he was much more explicit than Bohm, for not only does he provide us with a mechanism for such interaction (the inner octave), by the very nature of the octave itself he further presents an entirely cohesive worldview expressed in terms of exact musical symmetries and proportions. And, according to Gurdjieff, these same symmetries and proportions are present in man because the individual is, in effect, a "miniature universe," what Bohm might call a holographic imprint of the deeper, ultra-implicate reality. Gurdjieff, however, then qualified this comparison by stating that a complete parallel between man and the world can only be drawn if we take *man* in the full sense of the word: "that is, a man whose inherent powers are fully developed. An undeveloped man, a man who has not completed the course of his evolution, cannot be taken as a complete picture or plan of the universe—he is an unfinished world."[2]

In quantum terms we might say that such an individual has not yet acquired a nonlocal condition of "optimum psychological resonance" and so is unable to project psychokinesis influences out into the hierarchy of dimensions at a high enough or deep enough level. "Height" and "depth" are each seen in this context as properties of the greater, nonlocal reality, in the sense that the higher or finer vibrations—the inner octaves described by Gurdjieff and embodied in the Magic Square of Egyptian and Greek metaphysics—penetrate deep into the heart of everything.

For me, the most important aspect of "Gurdjieff's system" (he would never claim it as his own) is the way the individual's place in the cosmic scheme of things is so clearly defined. It seems that we all have a place in this worldview. In our case, this "place" is presently the planet of our origins. Significantly, however, Gurdjieff's system also provides us all with a purpose in life, one that offers a way of striking out into deepest space and enhancing the very presence of the planet on which we were born. Our raison d'être, he said, is to evolve, to develop and expand our consciousness to the degrees of resonance at which it can encompass these higher dimensions, way beyond the scale of planet Earth and the solar system, and even the galaxy. Of course, as I have suggested several times, the fact that this system is based, like DNA and the genetic code, on musical principles and symmetries means that this kind of "spiritual" growth—the development of human consciousness from the scale of its origins up to a greater scale above—is, like all creative processes, fundamentally an organic mode of evolution.

Gurdjieff said that ordinary "socialized" human beings are little more than complex machines, automata, living under the forty-eight orders of laws governing life on Earth (world order 48 in the "ray of creation"), constantly reacting, mostly involuntarily, to external stimuli. The laws and forces governing each of the worlds in the ray of creation, he said, are entirely mechanical, manifesting and interacting strictly according to the law of triple creation. So the evolution of the human psyche, or the development of what yogis and mystics call "cosmic consciousness," is seen here as a metaphysical journey up through the higher worlds and dimensions, at each stage of which the individual frees him or herself from a certain and definite number of the prevailing laws and forces of the particular world order in which they exist. For example, according to Gurdjieff, we on Earth are separated from the "Absolute," or the ultimate scale, by forty-eight mechanical laws. If we could free ourselves from one half of these laws we would be one stage nearer to the Absolute scale of existence and subject only to the twenty-four laws governing the next world order—the overall planetary sphere. Again, freeing ourselves from half of these laws would gain us access to the next world, the sphere of the sun or the solar system, where we would be subject to only twelve mechanical laws—and so on, with six laws controlling the world of "all suns," that is all solar helices, and three laws, three fundamental forces, controlling the greater world of the galactic helix.

This familiar description of the natural process of transcendental evolution embodies the essence of Gurdjieff's system of self-development, which was designed specifically to assist his students in systematically freeing themselves from these mechanical laws. No "miracle," he said (by which he meant psychokinesis, telepathy, and so forth) occurs as a result of the violation of these laws; a miracle can only be a manifestation of the laws and forces of a higher world.

Obviously the scientific community in general is opposed to the idea that the mind can engage in paranormal activities: it requires evidence that is measurable in some way. Shamans, yogis, mystics, and teachers of esoteric wisdom, however, do not. They appear to "measure" things, phenomena, experiences in a very different way from the modern scientist; that is, they assess and comprehend nature not only logically, with their minds, but holistically, that is with their whole being.

Ouspensky recognized the difficulty in observing the paranormal by purely scientific means after a period during which he experienced a number of telepathic encounters with Gurdjieff. These occurred during and after a field trip to Finland with Gurdjieff and a small group of his students a short time before the Bolshevik uprising. Just prior to this, Ouspensky had been taking part in a series of rigorous mental exercises and short but intensive fasts, which induced in him an unusually excited and nervous state.

One evening Gurdjieff called Ouspensky and two others to sit with him in a small room of the country house in which they were staying. Gurdjieff proceeded to show them some physical movements and postures, after which he gave them a brief talk on certain matters recently under discussion. It was at this point that Ouspensky had an experience he would never forget.

It all started with him beginning to hear Gurdjieff's thoughts. He said that Gurdjieff was talking to those present in the normal way, when suddenly he noticed that among the words Gurdjieff was saying were separate "thoughts" that were intended for him alone: "After a while I heard his voice inside me as if it were in the chest near the heart. He put a definite question to me. I looked at him; he was sitting and smiling. His question provoked in me a very strong emotion. But I answered him in the affirmative."[3]

To the obvious astonishment of the other two present, this intermittent "conversation" lasted for about half an hour, with Gurdjieff posing questions silently and Ouspensky replying in his natural voice. The substance of this dialogue Ouspensky declines to detail, but it seems that the questions posed by Gurdjieff were very difficult and sometimes of an extremely personal nature. Eventually Ouspensky became so agitated and disturbed by the proceedings that he hurried out of the room and escaped into the surrounding forest to try to gather his thoughts.

When he returned to the house it was dark. Unaware that Gurdjieff and the others were having supper on the veranda and thinking everyone had retired for the evening, he went to bed. But then, after a while, he began to feel a strange excitement and his pulse began to beat forcibly. At this point he once again heard Gurdjieff's voice inside his chest.

This time, however, he was able to reply to the question mentally and it seems that Gurdjieff "heard" and responded.

Much to Ouspensky's obvious discomfort, this extraordinary state of affairs continued for several days. Eventually the group traveled back from Finland to St. Petersburg and then met at the main railway station to see Gurdjieff off on a train bound for Moscow. Ouspensky then reports, "But the miraculous was still far from ended. There were new and very strange phenomena again late in the evening of that day and I "conversed" with him while seeing him in the compartment of the train going to Moscow."[4]

You can make of this what you will. Ouspensky, as he reports in his book, experienced other unusual states of awareness at this time, some of which, as he himself admits, he may have imagined. But when speaking of these extremely lucid telepathic encounters with Gurdjieff, his account is quite precise and unequivocal. As far as Ouspensky was concerned, he was communicating with Gurdjieff through an entirely different and much more efficient mode of transmission than ordinary vocal means.

So, if telepathy is a reality, how does it work? The answer currently on offer is, of course, "waves and interference patterns," though of a kind far removed from the wavelike form of subatomic quanta, of interactive electrons, packets of light, and all the other subatomic paraphernalia currently haunting the nonlocal world of local scientists.

The problem, as Ouspensky saw it, is that "metaphysical" phenomena cannot be investigated by ordinary methods:

> It is a complete absurdity to think that it is possible to study phenomena of a higher order like "telepathy," "clairvoyance," foreseeing the future, mediumistic phenomena and so on, in the same way as electrical, chemical, or meteorological phenomena are studied. There is something in phenomena of a higher order which requires a particular emotional state for their observation and study. And this excludes any possibility of "properly conducted" laboratory experiments and observations.[5]

As we shall see, the reference here to the emotional state of the investigator has a significant place in Gurdjieff's interpretation of the theory of transcendental evolution. The point to note here is that the paranormal aspects of the human psyche described above were considered by Ouspensky, a scholarly and rather stoic Russian intellectual, as phenomena deserving of study, that is he believed them to exist.

Probably very few of us living in the modern world will ever experience or witness real paranormal happenings. But of course, we all inhabit a predominantly secular, "socialized" environment controlled outwardly by legislation and underpinned by an economic substructure whose material demands upon us leave little time for voyages "into the mystic." This does not mean, of course, that the mystic is simply a figment of mankind's collective imagination. In fact, as Michael Talbot points out in his book *The Holographic Universe,* the evidence for paranormal psychic abilities as manifested through thousands of individuals in history is too compelling not be taken seriously. The number of serious researchers who believe that the holographic model can explain virtually all such phenomena is growing steadily.

The psychiatric researcher Dr. Stanislav Grof, for example, who spent several years studying the effects of LSD on thousands of volunteers, believes that the essential features of "transpersonal experiences" ("trips"), such as the sensation that there are no boundaries, no separate, unconnected elements, no distinction between parts and the whole, are all details one would observe in a holographic universe. He also thinks that the enfolded nature of space and time in the implicate realm is responsible for the feeling of timelessness experienced by so many of his volunteers. Unusual states of consciousness, he believes, can penetrate through to the implicate, enfolded order of things and modify phenomena in the physical world—the "images"—by "influencing their generative matrix."[6] Grof is saying, in effect, that the mind, as well as being capable of moving objects through psychokinesis, may also, under the right circumstances, be capable of influencing the source of these phenomenal images—what Talbot calls the "cosmic motion picture projector"—and so remodel the material world into any desired shape or form.

LSD—lysergic acid diethylamide—was discovered quite by accident in the late 1950s in Switzerland, synthesized from a fungus that forms on rye grain. Its effects on the human psyche were subsequently realized and explored extensively by the psychedelic generation of the 1960s.

Several decades prior to this, Ouspensky was also undergoing some very interesting "transpersonal experiences" of his own, induced by some other form of mind-altering agent, possibly nitrous oxide, or "laughing gas." In the context of our investigation into the nonlocal world of the quantum psychologist, his recollections of those experiences are highly significant. This is what he has to say about these experiences in his book *A New Model of the Universe:* "The new world with which one comes into contact has no sides, so that it is impossible to describe first one side and then the other. All of it is visible at every point, but how in fact to describe anything in these conditions—that question I could not answer."[7]

Clearly this description is very similar to the observations made by Grof's volunteers, of a nonlocal world in which there are no boundaries, no distinctions between parts and the whole, and everything is wholly visible from any point of reference.

As Ouspensky discovered, the problem with short cuts like psychedelic excursions is that they are very intense and sometimes traumatic—so much so that the unprepared mind is often incapable of remembering even the essence of its experience, let alone competently recording it. Ouspensky himself said that during these states of altered awareness he found it impossible to finish a simple sentence because, between words, so many relevant and interconnected impressions came to him that he simply couldn't keep track of events in the normal way: ". . . and these new and unexpected experiences came upon me and flashed by so quickly, that I could not find words, could not find forms of speech, could not find concepts, which would enable me to remember what had occurred even for myself, still less to convey it to anyone else."[8]

Obviously these kinds of chemically induced perceptions of differ-ent realities are of limited value. They are stolen glimpses, so to speak, taken for the most part by people whose mental powers are not suf-ficiently developed and whose understanding is consequently limited. Possibly this is why subjects under the influence of psychedelics are invariably rendered speechless, unable to describe even a small part of their experience. Indeed, being unprepared, many people have been shocked and even frightened by the things they have seen. Doubtless, for every hippie who can still remember his or her "transpersonal expe-riences," there will be several that cannot remember, or don't want to.

Another interesting description of the wider reality, one that again encompasses the kind of nonlocal dimension we are investigating here, is provided by Paramhansa Yogananda in his earlier-mentioned book *Autobiography of a Yogi* (1946). Unlike Ouspensky, Yogananda probably experienced these perceptions not as a result of the use of psychedelic agents, but rather through a heightened state of awareness induced by extensive yogic exercises—in breathing, posture, meditation, and so on. In the chapter entitled "The Law of Miracles," Yogananda discusses in some detail his observations concerning the phenomenon of light and its place in the cosmic scheme of things. The passage quoted here, as one can see, is remarkably prophetic and actually reads, in detail if not in style, like a page from Michael Talbot's book:

> Motion pictures, with their lifelike images, illustrate many truths concerning creation. The Cosmic Director has written His own plays and has summoned the tremendous casts for the pageant of the centuries. From the dark booth of eternity He sends His beams of light through the films of successive ages, and pictures are thrown on the backdrop of space.[9]

Remember that this particular "holographic" metaphor appeared in print decades before the hologram was ever dreamed of. He continues:

Just as cinematic images appear to be real but are only a combination of light and shade, so is the universal variety a delusive seeming. The planetary spheres, with their countless forms of life, are nought but figures in a cosmic motion picture. Temporarily true to man's five sense perceptions, the transitory scenes are cast on the screen of human consciousness by the infinite creative beam. A cinema audience may look up and see that all screen images are appearing through the instrumentality of one imageless beam of light. The colorful universal drama is similarly issuing from the single white light of a Cosmic Source. With inconceivable ingenuity God is staging "super-colossal" entertainment for His children, making them actors as well as audience in His planetary theatre.[10]

Like Bohm, the Yogi sees the psyche of the observer as an integral part of the entire "cosmic motion picture." As Yogananda himself saw it, we are all of us legitimate members of the cast of actors in this universal pageant, performing images, cast onto the "backdrop of space," acting out plays within a play, as real and as unreal as the images on a cinema screen.

Significantly, we are here reminded of the Hindu belief that the mechanism that facilitates this great cosmic spectacle is in fact light itself—the "single white light of a Cosmic Source." As we noted previously, this is emphasized in the principal annual festival of the Hindus, known as Diwali—the festival of light. Yogananda said that Indian holy men, people who "know themselves," are able to "travel" at the speed of light and utilize the "creative light rays" to bring into visibility any physical manifestation. The "law of miracles," he said, is operable by any man who has realized that the essence of creation is light. This, he asserted, was the secret of psychokinesis, telepathy, and so on, powers that manifest as a result of tuning in the vibrations of the mind with the vibrations of light. Presumably such a meeting of forces would create a whole new web of interference patterns (inner octaves) and with it a whole new range of phenomena.

15

QP2: The Universal Paradigm

As we noted previously, the holographic principle—of the whole existing in every part—can be expressed quite simply in terms of musical theory, where each tone, semitone, quarter-tone, and the like of a major scale contains within it all the data needed to recreate the scale in full. We noted in chapter 8 that the principle of nonlocality is also accommodated by musical theory, in that the ultimate note at the top of a given octave, being also potentially the first note of the next scale, can exist in two places/scales at one and the same instant. In the same way, the top note of a given triple octave would exist in four scales simultaneously. And so on.

Now, the two fundamental musical laws of nature embodied within *pi*—the law of three and the law of octaves—tell us that all human beings are "triple octaves" of resonance, walking "trinities" composed of our sensations, emotions, and perceptions. In the Book of Revelation, which is one of the most detailed and revealing hermetic texts in existence, this internal trinity is symbolized by the image of the Woman in Heaven, the pregnant (fully realized) goddess of the skies, whose "child," after birth, becomes united with God. Significantly, she is associated with three distinct sources of cosmic radiation emanating from above, namely the stars, the planets (symbolized by the moon), and the sun.[1]

Assuming that these three "octaves" of the human psyche are harmoniously composed, a transcendental, twenty-second "note" (the "child" referred to in the above cosmological description by St. John) is then created. This is an entirely new phenomenon capable of existing in a greater scale, a higher dimension. We came across this extraordinary creation in earlier chapters: it is the concept, the "immaculate conception" of Christian tradition.

The Hermetic Code, remember, as well as embodying within it the holographic principle of inner octaves and of nonlocality, is also in essence a description of an organic process of evolution. This is to say that the individual's relationship with the greater cosmos is "biometaphysical," a repetition, on a higher scale, of DNA's relationship with its host organism. We can thus regard the concept as being the metaphysical equivalent of the amino acid. As I explained previously, the amino acid is the transcendental product of the RNA triplet-codon template, which is composed from three of the four fundamental nitrogenous bases. In precisely the same way, the concept, derived from the harmonious combination of three fundamental "metaphysical bases"—sensation, emotion, perception—must also be a transcendental phenomenon, an active, organic component in a higher, more complex process of development.

If we note that the Book of Revelation associates these three fundamental forms of human impulse with the forces operating in the trinity above—that is the stars, the planets, and the sun—we can depict the structure of the human psyche and its relationship with the greater cosmos like so:

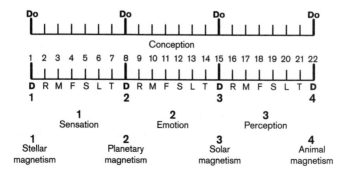

As we can see, the entire "composition" of the human condition and its relationship with the higher dimensions is described in perfect detail by the Hermetic Code. The influence of the stars, the planets, and the sun in the diagram above, I have called, for want of a better word, magnetic, although the term orthodox scientists generally abhor—*cosmic*—might serve just as well. The fourth metaphysical base, animal magnetism, is an internal power source, and from it is derived the concept, the product of the harmonious combination of the first three "bases," represented in the diagram as the ultimate note of the scale.

The idea that "cosmic vibrations" emanating from the spheres above might in some mysterious way influence the human psyche has always been anathema to modern science. Many scientists today, I am sure, would squirm at such a possibility. And yet, in the light of the universal hermetic processes referred to in practically every chapter of this book, the orthodox scientific view appears completely untenable. Just think about it. Are we seriously to believe that there is no connection whatsoever between ourselves and the greater cosmos, no continuity other than the purely chemical connection between the atoms of which we are composed and the exploding supernovae from which they originated? As we have seen, hermetic theory, a truly universal theory, emphatically excludes this kind of isolationism, and it even provides us with a coherent theoretical basis for such an exclusion. We have already seen this universal/hermetic connection in the way the seven dimensions of the universe coexist and interpenetrate, and how the four "base notes" of the entire evolving cosmos—DNA, the human brain, the solar helix, and the galactic helix—each function on their own scale, quite literally, as active, organic nuclei in the cells of the bodies of greater beings.

In Revelation, the four "base notes" of the trinity within us all (sensation, emotion, perception, conception), which together represent a microcosmic replica of the greater universal being, are symbolized by the four "beasts" sitting at the foot of God's throne—the lion, the calf, the man, and the magical fourth, the "flying" (transcendental) eagle. The eagle, in this context, is the concept, the product of animal magnetism. This is fire, the fourth "rare earth" of the alchemists, a human condition that Gurdjieff would describe as the law-conformable product of a real and independent will.

So, from the harmonious combination of the three principal "octaves" of the human psyche is born the last fundamental "note" of the whole scale, the transcendental fourth "base," out of which proceeds our fourth faculty, our ability to "immaculately conceive," to formulate enduring concepts and, no doubt, to perceive the world with a much greater degree of understanding. Representing as it does the ultimate note in the scale of human evolution, animal magnetism has qualities unique to it alone. It contains, at one and the same time, all the vibrations contained within the twenty-one notes below it. Furthermore, once this final link in the evolutionary chain of events is set in place, the whole phenomenon—in this case the psyche of the individual—resonates within as a unified whole, so that each of the twenty-two separate components of the given triple octave, being harmoniously related in the fullest sense, has nonlocal/holographic properties and contains within it all the information needed to recreate all three scales.

Taking the principle one step further, we can say that the whole psyche of the fully developed individual, all three inner octaves, also represents a single new note in a greater macrocosmic scale existing far beyond the confines of the brain. And if this greater scale of cosmic resonance is also harmoniously composed, then the higher "notes" comprising it would also be integral parts of a unified whole. Therefore each note of this greater scale (one of which, remember, is the human psyche itself) would possess nonlocal or holographic attributes and would therefore contain within it all the information contained within the whole.

As we see, hermetic theory not only supports the notion that the psyche is holographic within, it also accords with Bohm's view that it is an interactive part of a greater holographic process operating "out there." As Gurdjieff said, we are miniature universes, and within us are exactly the same laws and forces operating in the greater universe. It is this very fundamental truth that the enigmatic Egyptian "scribe of the gods" intended to convey to mankind when he first expressed the universal paradigm, "As above, so below."

Over the last few thousand years of human evolution, this concept of universal compatibility—this genuine "immaculate conception"—has persisted in the traditions of civilization builders across the entire world. Heaven is above, Earth is below, and the former dimension is accessible to anyone living what preachers of religions call a "righteous" life, that is, a way of being that reflects, and is compatible with, the prevailing laws and forces of the higher world, of heaven.

So, what would it mean for an individual to achieve this kind of cosmic consciousness? What might be going on inside the head of such a human being?

In an earlier chapter I discussed the work of Robert Jahn and Brenda Dunne, who conducted an extensive series of laboratory-controlled experiments to test the psychokinetic abilities of ordinary volunteers. Having concluded from their findings that such powers, albeit slight, do in fact exist, they have proffered an explanation for them. Basically they believe that consciousness itself, like all physical processes, possesses a kind of particle/wave duality, and that, when it is in a wavelike mode it can produce effects at a distance, for example psychokinesis or telepathy. If Ouspensky's previously mentioned telepathic encounters with Gurdjieff operated in such a fashion, that is through interpenetrating waves of inner octaves, or interference patterns, this means that such effects can span tens and hundreds of miles, and even lock on to a moving train.

The obvious implications of the proposition put forward by Jahn and Dunne is that the whole brain, when functioning at these higher levels of awareness, begins to behave something like a single quantum or wave/particle, whereby vast numbers of neurons occupying a given region of the cortex simultaneously act in concert, resonating at a frequency common to all the others—a kind of microcosmic version of what Colin Wilson calls the "group-mind" phenomenon, whose collective power greatly exceeds the sum of its parts. As I mentioned in chapter 4, the Cambridge mathematician Roger Penrose has come surprisingly close to this idea by suggesting that "non-local quantum correlations" between fundamental particles might be involved in conscious thought processes activated in the brain, so that a greater degree of awareness is possibly a direct result of this kind of "highly coherent quantum state." So perhaps neurons themselves can be induced to behave like particles, engaging in nonlocal quantum correspondence with other neurons and working as a single entity, as they seem to do in the case of the simple sponge.

We now have plenty of evidence of this "group-mind" activity in the quantum world, between electrons in plasmas, for instance, and in the phenomenal, instantaneous alignment of millions of molecules in the thermal process of formation of the Bénard cell. Another interesting example worth considering is the phenomenon of superconductivity.

A superconductor is the name given to any material that can carry a measured current of electricity with absolutely zero resistance. Normal conductors, such as the copper wires of electrical appliances in the main circuit of your home, invariably put up a certain amount of resistance to the flow of the current. This is usually lost along the length of wire in the form of heat energy. Under normal temperatures, known superconducting materials behave in exactly the same way, partly resisting the flow of the current and losing energy through a dispersal of heat. When they are cooled to temperatures approaching absolute zero, however, at a certain stage there occurs a magical transition, and the whole conductor seems suddenly to switch over to an overall "coherent quantum state" with absolutely zero resistance. It is as if every individual electron has suddenly transcended to a higher level of concerted "awareness," so that the whole conducting surface begins to perform like a single giant electron.

I'm suggesting here that something very like this kind of "super-unification" may also occur in selected regions of the brain, where individual neurons, given the right kind of impetus, can also transcend the ordinary physical world of cause and effect and unite as a single giant neuron with extra-cortical or transcendental properties.

But what kind of impetus is required to induce in the brain a highly coherent quantum state? Curiously enough, superconductivity may provide us with a clue here. We shall see how in a moment.

We are now assuming, of course, that higher states of awareness are indeed a reality. Such states have been spoken of not only by respected illuminati like Gurdjieff and Ouspensky, and prominent Indian yogi masters like Yogananda and Sri Aurobindo, but also, as we have seen, by a number of forward-thinking scientists in disciplines physical and metaphysical, ranging across a wide spectrum of intellectual effort, from quantum mechanics to clinical psychology.

As we know, yogis and other spiritual masters say that the higher psychological frequencies required to pierce the veil of space and time are accessed through work on oneself; through disciplined exercises in posture, meditation, and contemplation; and other specialized activities. Clearly, therefore, meditation, a process of stilling the mind, is a key factor.

To still the mind, as I have explained at length in *The Infinite Harmony,* is to switch over from an active mode of thought into a passive mode. Now this process is not dissimilar, in principle, to the cooling down of the superconductor, which brings about a drastic reduction in particle activity within its atomic structure. In effect, the superconductor becomes passive, and when it does the miraculous transformation in its conductivity occurs and there is suddenly absolutely zero resistance to the flow of the current. Something very like this, I believe, is what happens inside the brain when it is stilled to a sufficient degree; it turns into a kind of metaphysical superconductor, in which mode it puts up absolutely zero resistance to the flow of the "cosmic current."

In scripture, this process of switching over from active to passive mode is generally expressed through the idea of offering some kind of "sacrifice to God," that is to the higher vibrations of this world. It might be a very personal sacrifice, like giving up something dear to oneself, maybe a habit or tendency (as in Lent or Ramadan), or perhaps some time (through prayer or contemplation), or even something as seemingly mundane as your last-but-one goat, which at the start of the Muslim festival of Eid is traditionally killed and distributed to the needy.

At its heart, this sacrificial element is, in my view, the key to the whole process of transcendental evolution. It is spoken of in many ancient esoteric traditions, and I believe it refers principally to a psychological frame of mind in which we as individuals give up our preoccupations with secular trivia and switch over into an overall passive state. And this, of course, is precisely why in religion so much emphasis is placed on meditation or contemplation. Only a passive, receptive mind can take in active data, as a superconductor takes in an electric current, and subsequently transmit this energy, or whatever, in the most efficient way possible.

Significantly, this fundamental requirement—that is, that the mind must be passive in order to conceive, or to create—is graphically emphasized in Revelation, through the symbolism of the Woman in Heaven, a feminine, passive, receptive entity, who duly conceives and subsequently gives birth to a transcendental phenomenon—the Holy Child, the personification of the "immaculate conception."

Gurdjieff expressed this concept quite clearly during discussions with his groups in Moscow and St. Petersburg. In one talk, for example, he mentions a certain book of aphorisms, a collection of home truths gathered from unnamed sages. The following quotations from it clearly emphasize the importance of the passive element in the evolutionary processes of the mind: "A man may be born, but in order to be born he must first die, and in order to die he must first awake."[2]

We can interpret this passage in the following way. A man may be "born": that is he may, as Gurdjieff would express it, begin the course of his evolution. In order to do this, he must first "die": he must sacrifice everything for the good of the work. In order to be able to sacrifice all for the good, he must first "awake": he must first realize the need for change in himself.

Another favored aphorism also gets right to the point: "When a man awakes he can die; when he dies he can be born."[3]

This is precisely what the alchemical process described in the ancient pyramid ritual was intended to express. To be reborn as a god, as an "Osiris," the initiate first has to "die," to become receptive, to open his "mouth," his mind. When this is done, consciousness, like all products of nature's organic processes, can continue to grow further and ultimately come to fruition.

As I have explained at length in *The Infinite Harmony,* the need for the properly evolving mind regularly to "take time out" is actually the basis of all major religious disciplines. This is why so much emphasis is placed on setting aside, at the end of a working day, or week, a period of rest, a sacred interval, a Sabbath. To this end, meditation, contemplation, devotional prayer, and acknowledgment of forces or powers greater than our own are all valid exercises in stilling the mind, in making it receptive to the unseen, evolutionary forces permeating the universe. At the root of all this "religious" activity, of course, is music, which is why the Sabbath has consistently been associated with the number 7, the seventh note, ti, of the major scale, from which is "born" the transcendental eighth note.

The concept of composing mind and body, as I have said before, is relatively straightforward in theory, but in practice, as anyone familiar with the subtleties of yoga will know, proper meditation, involving correct breathing, posture, and thoughts—correct everything—is a very difficult thing to sustain for prolonged periods of time. Yogis undergo years of disciplined training to achieve mastery over themselves in this way.

It is clear that Gurdjieff drew some of his ideas from Hindu philosophy, but his particular system of self-development was perhaps more pragmatic and better tailored to the systematic Western mind. In principle, however, the objective remains the same: the acquisition of spiritual and psychological harmony.

Gurdjieff always emphasized the fact that the human being is fundamentally a trinity within, a triple octave, possessing three "brains" or "centers"—the moving center, the emotional center, and the thinking center. He said that all three of these "brains" need exercise, need to be developed together, in concert with one another. If only the thinking center is developed at the expense of the other two, the result is an unbalanced individual—something like the absent-minded professor who can never find his keys or his umbrella. Alternatively, a highly developed physical "brain" coupled with undeveloped emotional and thinking centers would result in an equally unbalanced or disharmonious individual—the "beefcake" archetype; and so on.

These three fundamental archetypes Gurdjieff referred to as "man number one," "man number two," and "man number three." All had something to contribute toward the development of the whole, he said, but the alchemical formulas for getting the mix right were solely the property of "man number four"—the "sly" man—possessor of animal magnetism, an individual who, like the eagle beast of Revelation, could "fly" to places where the first three could not.

The "places" in question are, of course, the nonlocal spheres discussed in previous chapters, the higher dimensions existing beyond space and time: the fifth, the plane of light; the sixth, the "solid" world of ultimate reality; and the seventh, the medium in which all "solids," all six-dimensional entities, exist and operate.

As we have seen, the first of these higher spheres—the plane of light, or the quantum field—has been pretty well charted by scientists. So this alternative reality definitely does exist. If it did not, there would be no such thing as a photon quantum, no means by which to gauge the maximum velocity allowed by nature—the "constant" or speed of light—and the theory of Special Relativity, with its implications for the "elasticity" of space and time, would be meaningless. This concept may be difficult for us to understand fully with only our ordinary logical thought processes, but the fact is that science has proven beyond all reasonable doubt that the quantum picture is the primary reality, and that our ordinary perception of both space and time is simply, as Einstein phrased it, a "stubbornly persistent illusion."

Nothing new here, then: after all, mystics and yogis have been telling us this for centuries, that there is no space, no time, no separate, isolated "things," that heaven is eternal and its extremities infinite. And all of these observations, as we have seen, are in accord with the quantum picture of reality.

These days, of course, established religious disciplines, originally designed for the express purpose of inducing in devotees the kind of altered states of consciousness we have been describing, are for many people decidedly passé. Others regard religion as little more than archaic mumbo jumbo, the "opium of the people," while the more vehement critics will argue that it has ultimately been the root cause of more murder, war, and bloodshed than any other human invention—as if man's inherent selfishness and his insatiable appetite for wealth and power had nothing to do with it.

Now there are, as we have seen, shortcuts to these higher planes of consciousness, such as the use of chemical triggers like nitrous oxide and LSD. Stanislav Grof's research led him to conclude that psychedelic trips—transpersonal experiences—are in fact voyages into the quantum or subquantum field. Presumably there will also be a few million former hippies out there who can remember experiencing the sensations of timelessness and oneness. And if their lysergic acid was pure enough, and their constitutions strong enough, they might also remember having the distinct impression, as Yogananda did, that all material things are actually "undifferentiated masses of light."

Obviously, however, the legitimate way of perceiving one or another aspect of the wider reality is, quite simply, to work on oneself, to develop one's powers of cognition, and thus evolve. Gurdjieff said that work of this kind was best performed in a school situation, in which each individual member is able to assist, and can be assisted, in the execution of what he called their "conscious labors," that is labors done in the right frame of mind and spirit. There is thus a subtle but important difference between ordinary work and work for the sake of the Work. As Gurdjieff put it, there is more to be gained from simply sweeping a floor with the right intent than there is from writing a dozen books with the wrong one.

In one talk he likened the situation of the pupil to that of a prisoner locked in a cell, whose only hope of escape is to enlist the help of others. One maybe fashions him a rope, another steals him a key, a third perhaps acts as a decoy, while a fourth sits discreetly outside the perimeter wall in the getaway car. Thus, individual development within a school is in fact a joint effort, a "group-mind" situation, in which all must contribute for the good of the whole. Of course, once our escapee is free from his "prison," there is opened up a whole new range of possibilities for the comrades he left behind, for they now have additional "outside" help.

And so, what of the possibility of raising the level of individual consciousness? If we judiciously exclude the somewhat controversial method of ingesting psychedelics, or of being fortunate enough to locate an authentic school of self-development (beware of imitations), then it would seem, on the face of things, that there are precious few possibilities for us to perceive directly the kind of nonlocal reality described by physicists. But in fact this is not necessarily so. As I said when discussing Robert Bauval and Adrian Gilbert's description of the so-called pyramid ritual performed by the priests of ancient Egypt, this essentially "alchemical" or hermetic process can be reenacted by anyone, and one doesn't necessarily have to be inside the Great Pyramid for it to be effectively performed. Indeed, there are situations in life that can induce these receptive states of mind, nearly always bringing with them a markedly greater perception of the wider reality, a greater appreciation of the world about us.

For example, when in imminent danger of losing one's life, the brain, or its consciousness, appears to be capable of tuning in instantaneously to the quantum field. Thus, survivors of violent events like car, train, or aircraft crashes often speak of having "seen" their whole lives flash by them in an instant. This kind of panoramic perception, in which an entire lifetime is somehow condensed into a single brief instant, can only take place in a dimension outside ordinary time.

The writer Graham Greene, a manic-depressive in his earlier years, experienced something very similar when, in a moment of recklessness that would make ingesting psychedelics seem as harmless as taking tea with your grandmother, he picked up a revolver and proceeded to play Russian roulette. The hammer on the gun clicked, the realization that he was still alive dawned on him and, as if a veil had suddenly been lifted from his eyes, the world immediately appeared infinitely more rich and meaningful.

When consciousness suddenly "expands" in this way, the process appears to be very similar to what happens when the latent energy inside a microcosmic atom is released in a nuclear explosion and the resultant shock waves reverberate out into the macrocosm, or the atmosphere, in the form of a vast mushroom cloud. The physical brain housed in the skull is the "atom," a microcosm, a localized center of energy; consciousness itself, when operating at the kind of frequencies triggered, say, by imminent danger, is a macrocosmic manifestation of the selfsame energy.

Now there are, I believe, more gentle and amenable ways to expand consciousness, to pierce beyond the veil of ordinary time and space. For example, even something as simple as a hard-earned vacation can lift the spirits and make one noticeably more appreciative of the vast richness and variety of the world about us. In such situations, time can seem to fly by, whereas in duller moments we say it drags.

But in reality, of course, the world itself does not alter in any fundamental way; it is always full of wonder, and often all that is needed is a different perspective, a change of scenery, for us to sense that this is so. We have all, at some stage in our lives, experienced negative thoughts and emotions, felt deflated, bound in time, locked inside a "miserable" day, inside an hour, inside a tiny moment. At such times we see practically nothing of the world about us. On the other hand, you might also remember how rose-tinted the world looked when you had just "broken up" for the summer holidays, or when you were young and first in love, or perhaps camping out under a tropical sky with the stars so close you felt you could reach up and grasp them. Such experiences as these are, in fact, genuinely magical, and if one takes time out from the humdrum grind of secular living to reflect upon one's own life, they can usually be remembered quite easily.

Unfortunately, however, despite these illuminating incidents, we generally live the major part of our lives only in the fourth dimension of time, a sphere of existence that, in respect to the higher dimensions, is a narrow, essentially linear and extremely restricted world. This is where ordinary consciousness exists, isolated, like a faint spot of warmth moving along an invisible wire. And if this wire, the line of time, is, as I have tried to explain in earlier chapters, a sort of "cross-section" of a greater plane, then during such events as transpersonal experiences, consciousness, like the energy of the atom bursting out in a nuclear explosion or, perhaps, the mass of a galaxy as it draws toward the threshold of the speed of light, would theoretically expand, stretching out laterally, "at right angles," to the directional flow of the line of time.

Significantly, Yogananda says much the same thing when speaking of masters who are able to perform supernatural feats, that they "have fulfilled the lawful condition; their mass is infinite."[4]

He says further:

The consciousness of a perfected yogi is effortlessly identified not with a narrow body but with a universal structure. Gravitation, whether the "force" of Newton or the Einsteinian "manifestation of inertia," is powerless to compel a master to exhibit the property of weight, the distinguishing gravitational condition of all material objects. He who knows himself as the omnipresent Spirit is subject no longer to the rigidities of a body in time and space.[5]

Quantum psychology in a nutshell.

Not surprisingly, the Hindu tradition of Yogananda is steeped in hermetic lore. Thus the individual is regarded, as in all religious systems, as a living trinity, comprising a physical, astral, and mental body—the equivalent of Gurdjieff's three "centers," the moving, the emotional, and the thinking.

In a short but remarkably perceptive book, *The Theory of Eternal Life,* Ouspensky's associate Rodney Collin, drawing from Gurdjieff's original ideas, tries to identify the possible nature of these three bodies.

The physical body, he says, is the one we are all familiar with, and is fundamentally cellular in nature. The second, astral body, or the "soul," which, he suggests, grows as a result of developing the emotional center, is basically a molecular manifestation, like, say, sound or scent. Being of a finer, more fluid form of materiality, the astral body has powers unobtainable by the physical body. For example, like sound or scent, it would be able to diffuse many times faster than the cellular body moves. Further, a cellular body moves only in a line, whereas a molecular body would be able to spread out simultaneously over a wide area, like an aroma. Significantly, such a presence would also be free of the force of gravity, a fact that reminds us of Yogananda's claim that such a force cannot affect those who "know themselves."

Collin then goes on to imagine human consciousness endowed with the properties of matter in a molecular state. It could, for example, be present in many places simultaneously. It could pass through walls, assume a whole host of different shapes, even enter inside other human beings. Like musk, it might "haunt" a place for several years; and if a molecular body the size of a human being possessed the metaphysical equivalent of the "pungency" of mercaptan, which retains its nature even when diluted in fifty trillion times its own volume of air, it could be simultaneously conscious of every hectare of land in an area the size of China.

As Collin notes, *The Tibetan Book of the Dead* refers to this astral or molecular body as the "desire body," one that, unlike the gross physical body, has the power to "go right through any rock-masses, hills, boulders, earth, houses, and Mount Meru itself."[6]

The passage quoted is addressed to the dead person, and it implies that the molecular or astral body can continue to exist after the physical body has expired. It continues: "Thou art actually endowed with the power of miraculous action. . . . Thou canst instantaneously arrive in whatever place thou wishest; thou hast the power of reaching there within the time which a man taketh to bend, or to stretch forth his hand."[7]

Remarkable as these powers might seem, however, they would still fall short of the real thing, in that the astral body constitutes only one third of the complete trinity, vastly more complex and energetic than the physical body, but compared to the mental body born of what Gurdjieff called the thinking center, it would still be relatively small and limited. This is why the quoted passage finishes with a warning: "These various powers of illusion and of shape-shifting desire not, desire not."[8]

As Collin says, both *The Tibetan Book of the Dead* and the Egyptian book of the same name, along with many other ancient texts, all suggest that at the death of the physical body this new body of molecular energy—the astral body—is born. Buddhists believe that this acts as the vehicle of consciousness in the interval between incarnations. In fact, in Tibetan, Egyptian, and Peruvian rituals, fresh food and drink were set aside in the belief that the smell or essence of it would nourish the soul of the dead person—a practice Collin sees as clear recognition of the fact that the physical composition of the "soul" is similar to scent, that it consists of matter in molecular state. He suggests that this molecular body, like everything else, has to be created, and that this is accomplished through a sustained accumulation of the finest energies produced by the physical organism in life. And in order to do this, as Gurdjieff said, individuals must first create in themselves a will of their own, one that would empower them with the inner strength to restrain the wasting of these energies through the usual negative emotions or impulses—anger, fear, longing, envy, and so forth. This, according to Gurdjieff, is real alchemy, the "transmutation" of coarser energies (human emotions) into finer ones, or the transformation from an ordinary "individual" into one who is genuinely undivided, and whose inherent willpower is consequently fully developed.

In chapter 4, Collin goes on to speculate on the possible form and function of the third and last component in the human trinity, the harmonious product of the thinking center, that is the mental or, as he calls it, the "electronic" body. His ideas here are particularly interesting, because they bring us right back to the quantum field of the physicist, the plane of light.

As he says, molecular vibrations, like sound, diffuse about a hundred times faster than physical bodies move, but "electronic" energy (by which he meant light, the photon quantum), radiates nearly a million times faster still. Thus a body composed of "electronic" matter, which Collin calls the "spirit," could travel instantaneously through three dimensions: that is along a line, like a cellular body, over an area, like scent or a sonic boom, and throughout an infinite volume of space, like the proverbial Holy Ghost.

Collin then tries to imagine what would happen if human consciousness were attached to an electronic device, for example a bright lamp in a room.

First, the center, or heart, of the body would be the incandescent filament of the lamp, but it would also include all the light emitted from it. A consciousness attached to a body of this nature would contain within it all the objects in the room: furniture, flowers, and plants, even its occupants. Consequently it would illuminate or be conscious of every object from all sides simultaneously. Everything in the room would, in a sense, become inner organs of this electronic entity, and everything happening would be happening inside it and would be sensed as its own life. Thus human consciousness attached to a body of light, in including all neighboring beings within itself, would share the nature of "God," in whom, it is said, all creatures exist and have their being. It is precisely this principle of joining together, says Collin, that lies at the root of both yoga, which means "union," and religion, which means, "reunion."

The Tibetan Book of the Dead describes this form of consciousness as the "Radiance of the Clear Light of Pure Reality." The dead person, assuming he possesses a "spirit," is addressed thus: "Thine own consciousness, shining, void and inseparable from the Great Body of Radiance, hath no birth, nor death, and is the immutable light."[9]

According to Collin's interpretation of Gurdjieff's system, in order to acquire a spirit, an electronic body, an individual must first develop a soul, a molecular body. This is done by concentrating all of one's molecular energy—one's emotional output—to this single goal. Once this is achieved, the next step would be to "infuse soul with spirit," which means, in effect, that individuals have to learn how to transmute molecular matter into "electronic" matter, that is, "To split the atom and release internally a degree of energy which only our own age can begin to measure."[10]

As we see, we once again find ourselves being drawn into the quantum world inside the atom, the nonlocal realm of electromagnetic radiation, of the photon quantum, the Holy Ghost, the Immutable Light of Tibetan Buddhism.

Inside the atom, the nuclear forces ensure that the constituent particles are contained within definite energy levels. Thus the atom, together with its nucleus, is a hard nut to crack. Gravity also plays a part in it, though very small, as all matter exerts a gravitational pull, however slight.

The atoms themselves are held together by the electromagnetic force, which binds all matter. If its influence were removed, all material stability would cease. Without the electromagnetic force, the chair you are sitting on, or the ground on which you stand, together with the body you are presently occupying, would dissipate, dissolve into invisible clouds of free-floating quanta. There would be no molecules, inorganic or organic, only a homogenous soup of fundamental components—quarks—bouncing off one another in an endless display of randomness and non-interaction.

Now the force-carrier, or "gluon," of electromagnetic radiation is, of course, the ever-constant photon, the most special of all wave/particles. It is unique primarily because it is the only phenomenon in existence capable of inducing in us visual sensation. In other words, it is the medium via which we receive most of our impressions or perceptions, the bridge, so to speak, between mind and the empirical world. But bridges are for crossing, and in the light of the evidence discussed in this chapter it would appear that this particular one has been success-fully encountered by many a free spirit, some of whom, as we have seen, have left us with some extremely lucid accounts of the extraordinary things they have witnessed on the "other side." With one voice, they speak of a miraculous, timeless, spaceless world, brimming with con-sciousness and light, pulsating with waves upon waves of pure energy.

Scientists say that mass and energy are in fact different manifes-tations of the same thing. This is significant, because consciousness, although we cannot clearly define the phenomenon, can reasonably be regarded as a manifestation of a subtle form of energy. And then we have the photon, the other side of our metaphysical coin, a "virtual" particle that, having practically no measurable mass, exists on the very edge of materiality. Moreover, of all particles known to exist, only the photon has no antimatter opposite; the photon is its own opposite. If you were to draw a line down the center of a piece of paper and list all particles on one side and all antiparticles on the other, the photon would have to be placed over the line. It is in every sense a duplicitous entity, a shapeshift-ing Jekyll and Hyde, a particle and a wave. Catch it if you can.

Thus we can see that the dividing line between mind and matter is in reality extremely tenuous, so much so that one feels it would not be stretching credibility too far to propose that a high degree of conscious-ness resonating "in tune" with photon quanta might somehow temporar-ily neutralize the electromagnetic force, thus making any form of matter present within the sphere of neutralization "fluid." It need only be a fleeting moment of fluidity, imperceptible to the naked eye, but if actu-alized repeatedly in short, sharp bursts, it could well be sufficient, say, effectively to reshape or move great chunks of stone with relative ease.

Such powers, in my view, are attainable, but I believe that they are simply a by-product of quantum psychology, the transcendental evo-lution of the mind. The Egyptians and the "builder gods" of ancient America, and probably their mysterious forebears, the original initiates, were clearly past masters of the art. It might be appropriate, therefore, to leave the last word to them.

16

The Shapeshifters

In this era of scientific rationalism, the subject of psychokinesis is, not surprisingly, a virtual nonstarter. Even though we frequently hear or read about individuals who have, or have had, supernatural powers, very few people have actually seen, or even laid claim to seeing, proof of such things. So if we exclude party tricks, table-turning, spoon-bending, fakirs on beds of nails, and other such forms of light entertainment, there is really very little in the way of what we might call the genuine paranormal in the day-to-day life of modern man. But possibly this is simply a sign of the times: the overt rationalism of science over the last few centuries may have dulled our extrasensory abilities almost to the point of atrophy. As Ouspensky said, phenomena of a higher order seem to require a certain degree of emotional energy for their observation and study, and this would automatically preclude ordinary experimental methods. This is not to say, of course, that scientists as a whole are unfeeling, only that experimental observers' personal level of inner development, including their emotional state, is not yet considered an essential factor in objective scientific investigation.

Paradoxically, however, quantum physics, the quintessential science of rationalists, is a veritable hotbed of the paranormal, with its non-locality, superconductivity, and the peculiar phenomenon known as "quantum tunneling," which involves specterlike virtual particles popping up everywhere, out of nowhere, only to disappear again nanoseconds later, leaving no measurable trace.

In the quantum world, practically anything is possible, even if highly improbable, so why not so in the meso- and macrocosmic worlds, where everything is, after all, made up of quantum systems, and where exactly the same laws of physics apply?

Obviously individual quantum effects are too small to leave observable traces on ordinary sense-objects. A chair remains pretty much the same chair no matter how many "nonlocal" photons are absorbed or reflected by it. If the chair behaved like a single microcosmic entity—like an atom of some kind—things might be very different. It could be a chair one moment, a shimmering mass of waves and interference patterns the next, or perhaps both things at once. Alternatively, if our atom-chair were to interact with a single photon, it might suddenly change into another kind of chair, one of a different color perhaps, or of a different materiality.

We noted previously how superconductivity belies experimental logic: the conductor may suddenly and inexplicably change its state entirely. Under normal conditions a superconductor is simply a mass of typical quanta: protons, neutrons, and electrons; whereas in its supercooled state, the whole entity subsequently begins to behave like a single unified system with quite extraordinary properties. So superconductivity is in fact a paranormal phenomenon. The conductor itself, although made of exactly the same stuff as you or me, is a conduit, leading, quite literally, to a higher dimension, in which a given electric current, in meeting absolutely zero resistance—the ultimate passive state—can potentially flow on to infinity.

Admittedly this kind of observation still leaves us very little to go on in our quest for evidence of psychic powers. But if today the paranormal is conspicuous principally by its absence, this does not seem to have been the case in ages long past. The scriptures, for example, by which I mean the major bodies of religious and esoteric wisdom, all speak of the miraculous as if, at one time, it was almost commonplace. Similarly, the legends and myths surrounding the great civilizing heroes of the dawn of history—Osiris and Thoth in Egypt; Viracocha, Quetzalcoatl, and Kukulkan in the Americas; the Aryan Manu, Fu-hsi of Chinese tradition; or Orpheus and Amphion in Greek legend—all feature these figures' supernatural powers. Half men, half gods, these civilizers could apparently move mountains. Certainly they must have been tough cookies to have survived the most geocataclysmic period in the history of *Homo sapiens sapiens,* not only the bitter climate of the last ice age, but also its final, rapid meltdown and the subsequent catastrophic flood.

This alone is a remarkable achievement, but this giant of a race not only outlived a seemingly endless hell on earth, they then went on to build like giants, leaving a wealth of archeological evidence as proof of their still unexplained mastery of stoneworking. Such individuals must have been the most organized, resilient, and resourceful ever to have existed. No time then for squabbles, fighting over possessions, petty neuroses, or living in the past. Life was serious, and for a person to have survived the deluge would have required, as a matter of course, nerves of steel, an unswerving will, a mind as clear as crystal. A quantum mind, perhaps, the kind that sees the world—and its pitfalls—non-locally: that is, from all sides at once, from "above," so to speak, from the plane of light, or even from the perspective of dimension six, the ultimate vantage point.

What dangers might such a mind face? Possibly only one: the Fall described in the Hebrew Scriptures and other ancient texts, the fall from grace, from overarching, "nonlocal" consciousness to the ordinary time-laden psychological currency of today.

Significantly, there is an old Native American text indicating that the great civilizers definitely did see things in a way very different from our own. This passage comes from the Popul Vuh, the sacred book of the Maya. The italics are my own:

They saw and *instantly* they could see *far;* they succeeded in seeing; they succeeded in knowing all that there is in the world. The things hidden in the distance they saw *without first having to move.* Great was their wisdom; their sight *reached* to the forests, the rocks, the lakes, the seas, the mountains, and the valleys.[1]

This describes, quite clearly, perception of a nonlocal kind, a mind that "sees" at a great distance as if there were no intervening space. Their sight "reached to" everything, as if from all sides at once.

Upon first reading this text, I was immediately struck by the similarity between the kind of "sight" with which the Mayan gods were empowered, and the telepathic experiences of Ouspensky, during which, it may be recalled, he was not only able to communicate with Gurdjieff, but also simultaneously to "see" him sitting in his train compartment, heading for Moscow.

The text of the Popul Vuh continues, describing the mysterious fall from grace of this race of supermen, who had evidently displeased the gods. A divine order was duly proclaimed: "Let their sight reach only to that which is near; let them see only a little of the face of the earth."[2] And then: "Their eyes were covered and they could see only what was close; only that was clear to them."[3]

But before the Fall, these people, according to legend, could not only see non-locally, they also had awesome psychokinetic powers, induced through special forms of music, with which they were able to move gigantic blocks of stone as if they were made of polystyrene. There is no direct evidence to support the mythmakers' version of the methods used then, before even the wheel and the pulley were invented. Nevertheless, according to widespread archaeological evidence, ancient builders possessed a level of expertise that seems eerily out of step with the established history of architecture.

Remember, in what are probably the oldest known buildings in Egypt—the Sphinx and Valley Temples and the Oseirion temple near Abydos—the average size of the limestone blocks is a staggering two hundred tons. These enormous blocks have been positioned one on top of another with extreme precision, yet the Sphinx enclosure, confined within an area enclosed in natural bedrock from which the statue itself was carved, would have been too small to admit a large number of workmen during construction of the adjacent temples. It is now accepted by the more progressive members of the archaeological community—at least in private—that the Oseirion and the temples of the Sphinx enclosure may predate the Fourth Dynasty by a considerable length of time. It seems as if, the further back you go, the more impressive and baffling the buildings are. Nevertheless, the Egyptians of the Third and Fourth Dynasties, the successors of these mysterious masterbuilders, also displayed spectacular stone-handling skills.

GP The Great Pyramid speaks volumes, of course, with its sheer magnitude and precision. The finely dressed and precisely positioned granite blocks hauled high up into the King's Chamber each weigh upwards of thirty tons. The chamber's mathematical exactitude is also interesting, particularly in light of the musical theory of transcendental evolution. Being exactly twice as long as it is wide, it gives rise to the ratio 2:1, which is the ratio that defines the first and last notes of the octave. Another interesting fact is that the GP itself, as well as having the familiar classical *pi* relationship incorporated in its dimensions and proportions, also appears to embody the "Greek" value of phi, which naturally occurs in the relationship between the Great Pyramid's base and the length of its apothem or slope; that is, half the base length is in the ratio 1:1.618 with the length of the apothem. Phi, like *pi,* is an irrational number (1.61083); the geometrical proportion derived from it—known to the Greeks as the Golden Section—was considered by the Pythagoreans to have a particularly distinctive aesthetic quality, a visual harmony of great value when expressed in architecture. It was subsequently incorporated in the structure of the Parthenon, whose ruins even today inspire in us all a deep sense of beauty and harmonic proportion. The Greeks, of course, could feasibly have been working from a long-extant blueprint, once the property of the ancient Egyptians, and that, even then, if the Egyptians themselves are to be believed, was a legacy from a much more remote era, called in the Pyramid Texts *Zep Tepi,* the "First Time."

A little later on we can take a closer look at this Golden Section, and we shall see that phi, like *pi,* has applications way beyond the parameters of geometry and architecture. We shall also see how, in light of the creation process described by the Hermetic Code, there is a subtle but significant connection between the two ratios.

It is now generally accepted that the dimensions of the so-called sarcophagus in the King's Chamber also incorporate exact mathematical values and proportions: an internal volume of 1,166.4 liters and an external volume of exactly double, 2,332.8 liters. Again, we can see here another example of the 2:1 ratio found in the structure of the ubiquitous octave. This remarkable artifact was apparently cut and hollowed from a single block of exceptionally hard granite with measurable precision. How did the stonemasons do this?

As we noted in chapter 3, the archaeologist William Flinders Petrie was impressed by the skill and expertise of the craftsmen responsible, but was at a loss to explain how the work was carried out. He surmised that the sarcophagus itself must have been cut out of the mother block with long straight saws, perhaps with bronze blades tipped with jewels harder than granite. However, as Petrie himself admitted, the only jewel hard enough to cut granite so efficiently would have been diamond, which, as far we are aware, was not known in Old Kingdom Egypt.

Petrie further puzzled over the method used to hollow out the sarcophagus, proposing, from the evidence, that hollow, tubular "drills" up to 12.5 cm in diameter were used to cut circular grooves in the granite, the cores of which could then have been chiseled away with relative ease. Again, these strange drills, powered by heaven knows what, are envisaged as having been jewel-tipped. After examining a number of other drill cores collected at Giza, Petrie estimated that the amount of pressure applied, shown by the speed at which the drills had evidently cut through the granite, would have required a load of at least one, or maybe two, tons.

Thus, according to Petrie, the ancient masons used tubular drills, their teeth tipped with something as hard as diamond and applied with pressures of up to two tons, revolving at speeds great enough to cut through granite as if it were as soft as soapstone. As to how such devices were set and kept in motion, Petrie could offer no explanation, but it would seem safe to assume that he had firmly ruled out pedal power. In any case, wheels and pulleys were supposedly unknown in ancient Egypt (an established archaeological "fact" that might seem hard to reconcile with the use of high-powered, circular drills). Somewhat mystifyingly, not a trace of any of these saws or drills has ever been found.

Petrie was equally perplexed by a number of Fourth Dynasty diorite bowls from Giza, on which, he observed, the engraved hieroglyphs had been drawn with a remarkably fluid, free-flowing hand, showing no signs of having been forcibly ground or scraped out. The lines of the inscriptions are extremely fine—a fiftieth of an inch wide—indicating that the cutting tool must have had a needle-sharp point. Bearing in mind that diorite is one of the hardest stones on earth, so that the point of the instrument used must have been a good deal harder—and again, supposedly applied with an enormous amount of pressure—it is difficult to explain how such fine motifs were so easily and smoothly inscribed. It is true that a modern tool could engrave designs on diorite with relative ease, but such a process would not leave rough edges to the lines. On the bowls identified by Petrie, however, there were such rough edges: the diorite had been "ploughed out," as if as soft as virgin soil.

Perhaps the most baffling examples of the Egyptian stonemason's art are the stone vases found in chambers underneath and around Zoser's Step Pyramid at Saqqarah, many of which date from centuries before the Fourth Dynasty. As I mentioned in chapter 3, Graham Hancock examined many of these anomalous artifacts: some of them had long, thin, elegant necks and widely flared interiors, often incorporating fully hollowed-out shoulders. As Hancock says, this kind of workmanship must have been accomplished by the use of some "as yet unimagined (and indeed almost unimaginable) tool."[5] Such an instrument would have to have been narrow enough to pass through the slender necks, strong enough to have scoured out the stone itself, while at the same time capable of exerting precise upward and outward pressure in order to shape the interior curves and shoulders. Hancock uses the word *scoured* to describe how these vases might have been hollowed out, but it would be interesting to see if the interior of the vessels, like the finely grooved hieroglyphs on the diorite bowls, was in fact "ploughed out," rather as if the stone, at some stage in the process, had temporarily become soft, pliable, fluid. But then, even if a tool for such a process had existed, this would still not explain how the craftsmen were able to gauge the progress and accuracy of their work, which proceeded, as it were, "in the dark," inside the stone.

The implications of all this are staggering, for we are not merely talking here of a stone-carving technology greatly in advance of its day, but one that is superior to our own. As Hancock says, no stone carver alive today would be able to match the workmanship of these stone vases, even if he were using the most advanced tungsten-carbide tools.

Speaking of which, whatever happened to all the tools that are supposed to have been used by these preeminent artisans? As we have noted, no traces of the long "bronze saws" have ever been unearthed in Egypt. The same applies to the hardened, tubular "drills" supposedly used to hollow out great blocks of granite. What is more, nothing approaching such devices has ever been depicted in any tomb relief; neither have they been described or even mentioned in any text.

And yet, the evidence that tools of this kind were used seems convincing. Examining one particular drill core, Petrie noted that the spiral of the cut sank one inch in a circumference of six inches, suggesting a rate of stone "ploughing-out" that he described as astonishing.

Many features of the Great Pyramid itself are equally extraordinary. It is extremely doubtful that even an army of modern stonemasons, using only ropes, levers, ramps, and handheld tools, could ever match such architectural mastery and precision. Measurements taken of its internal proportions and angles, of external alignments and dimensions, have been described as displaying a near-perfect symmetry. It should be remembered, however, that the Giza plateau suffered a violent earthquake in the thirteenth century. It is entirely conceivable, therefore, that prior to this the pyramid's overall symmetry might, in fact, have been even more precise.

In the BBC publication *Secrets of the Lost Empires,* Mark Lehner, described as one of the leading experts on the Giza Pyramids, says that the Old Kingdom masons used only copper chisels and punches, even to dress the harder granite and basalt blocks. His explanation as to how soft copper tools (Petrie was unaware that bronze was not in use in Egypt prior to 2185 BCE) could be used to cut and hollow out such incredibly hard stone is that the workmen probably used quartz sand in a wet slurry: it was actually the quartz that did the cutting. As evidence he cites the discovery of some cuts in the basalt of the Khufu/Cheops Mortuary Temple that still retain a dried mixture of quartz sand and gypsum, apparently tainted green from the copper saws. This proposition is almost, but to my mind not quite, feasible, for we would have to envisage a very large number of copper saws used in the cutting of even one megalithic block, as the quartz would have to have cut both ways, eating away the much softer copper tools a great deal faster than the granite was cut. Also, the presence of so much wet slurry in a hand-powered cutting process would surely have had a deleterious effect on the precision of the work. The near-perfect symmetry of the granite sarcophagus in the King's Chamber makes one wonder whether it really could have been achieved in this way. Furthermore, there is still the question of Petrie's estimation of the speed at which some of the granite drill cores had been cut, not to mention the extreme amount of pressure that must have been applied to the point of contact. As for how the engraved diorite bowls already discussed, and the remarkable stone vases described by Hancock, were produced, we have yet to hear of a plausible explanation.

Making no mention of these annoying little anomalies, Lehner describes how he conducted field experiments to demonstrate how he thinks the pyramids themselves were built. With a crew of forty-four, including a master stonemason, it was intended within a six-week time-scale to quarry and assemble 186 limestone blocks into a small pyramid, nine meters at the base and six meters high. Of course, modern construction equipment was employed in the operation. Flat-bed diesel trucks transported the blocks, weighing from three-quarters of a ton to three tons apiece, and a massive earth-mover was used as a crane to lift and maneuver them, slung from steel cables. In addition, the workmen used iron hammers, chisels, and levers, whereas the Egyptians, according to Lehner, had only copper tools and wooden levers.

In the event, Lehner and his colleagues did manage, within the six-week deadline, to construct a crudely dressed pyramid consisting of the specified number of blocks. This prompted Lehner to conclude confidently that the Pyramids of Giza, spectacular though they may be, were "very human monuments, created through long experience and tremendous skill, but without any kind of secret sophistication."[6] At no time did the project attempt to cut and dress granite blocks in the manner suggested by Lehner as that most likely to have been used by the Egyptians. It might be useful to see a practical demonstration as to how this was accomplished, not with iron or even bronze tools, but with the copper implements supposedly used by Khufu's masons. I find it hard to accept that relatively soft copper tools, even if used in conjunction with quartz sand in slurry, could ever cut efficiently through just about the hardest substances on Earth. Unless, that is, there was another, hidden factor.

According to Lehner, the Great Pyramid itself, a "very human" monument as he calls it, was built within Khufu's forty-year reign. So let's just think about this for a moment: forty years seems, on the face of it, a very long time for the construction of a single building. But, quite apart from the granite blocks, there are approximately 2,300,000 individual blocks of limestone incorporated in this structure, each weighing, on average, two and a half tons.

So, if all 2,300,000 of these blocks were quarried, transported, and set into position in the allotted time (40 years, or 14,600 days), then the rate of laying down—2,300,000 divided by 14,600—is 157 blocks per day. Dividing 157 by a working day of, say, 12 hours gives a construction schedule of 13 blocks every hour, one set in position every four and a half minutes or so, nonstop, that is incessantly, day in, day out, often in extreme heat, through all four seasons, for forty continuous years.

Already the incredible seems to have become a reality, but we haven't yet finished piling up the statistics. For example, it is unlikely that the building work was maintained at a constant rate for twelve months of the year. The annual flooding of the Nile would not only have hindered work on the Giza plateau, it would also have demanded that a substantial proportion of the available workforce be employed in the seasonal agricultural projects necessary to sustain themselves.

So let's reasonably assume that the builders worked constantly, not for twelve, but for eight months of the year or, alternatively, eight hours a day. We now have to envisage them laying down more than nineteen blocks every hour—about one every three minutes. But even then, if Lehner and his colleagues are correct in their assessments of the time involved and the volume of work done, we could quite reasonably halve this allotted time yet again. The reason? It is assumed that the limestone blocks (not to mention the thirty- to fifty-ton granite blocks situated high up inside the core masonry) were hauled up a wraparound spiral ramp built from gypsum and limestone chips—in mass almost equivalent to that of the pyramid itself. So, while the builders were busily hauling up and laying down the blocks, others were presumably adding millions of tons of extra material to the very same ramp in order to increase its height.

Remember there were no cranes, earth-movers, or diesel trucks, no cables, no iron tools—only ropes and wooden levers. As for lunch breaks, vacations, strikes, and all the other very human needs of a large workforce, there seems to have been little room left for such time-wasting in the life of the Fourth Dynasty pyramid builder.

Given all this remarkable statistical data, one is inevitably left wondering what kind of "human" enterprise we are considering here. It is not, in my view, one that could be matched today, muscle for muscle, stone for stone. Even if we doubled the proposed construction period to an unlikely eighty years, we would still have to envisage the blocks of the Great Pyramid being set into position at an unbelievable rate of, at the very least, nine or ten every single working hour. Lehner's team, if we include the accepted periods of respite, managed to lay down 186 blocks into the form of a very crude pyramid in six weeks, or 1,008 hours. This is the equivalent of laying down about one block every five hours or so, or one block every three hundred minutes. From the statistical evidence alone, we can see that the difference between the rate that the Egyptians laid down their blocks and that of Lehner's team makes the efforts of the latter seem nothing short of pathetic. Remember also that, unlike Lehner's team, the Egyptians had only copper tools and wooden levers, and supposedly had to build and maintain the massive spiral ramp up which the blocks were to be hauled. At the same time they needed to ensure that the pyramid itself, completely enveloped by this alleged ramp, was kept in perfect alignment, not only with the cardinal points, but also with the key stars above the Nile Delta. In addition, the workmen also had to ensure that the angle of each slope remained at 51 degrees, 51 minutes, which is the angle that would naturally result if the relation of the pyramid's height to the perimeter of its base was exactly the same as the ratio between the radius of a circle and its circumference.

In spite of its almost unbelievable level of precision and craftsmanship, we can still agree with Lehner on one fundamental point: the Great Pyramid is a very human monument. What he fails to emphasize, however, is that there are many possible degrees of "humanness," and that the degree of evolutionary development attained by the Fourth Dynasty Egyptians must, on the basis of the evidence they have left behind them, have been of an order in certain ways vastly more advanced than our own. We are not talking here simply about organizational skills and technological ability, but rather of the fundamental physical, emotional, and mental capacities of thousands, possibly tens of thousands, of highly developed, highly coordinated individuals, men who must have been in possession of long-lost powers of the human will.

Skeptics may argue, along with Lehner perhaps, that this is not so, that, given time, manpower, and finance—and the will, of course— modern man could replicate anything the Egyptians accomplished, even without modern equipment. But think again: this is a single building enterprise involving about six million tons of limestone and granite blocks, supposedly quarried entirely by hand, transported in barges, and then physically hauled up a narrow ramp, with the blocks being carefully positioned at the rate of several every working hour, eight hours daily, month in, year out, for four whole decades. I would suggest that modern man, if he were to attempt such a feat, would first have to rediscover certain skills and attributes that the Egyptians had and we definitely have not. As individuals, the people directly involved in the project must have been more than just physically fit: they must have been capable of sustaining their incredible physical output for very long periods, continually, incessantly.

So, exactly what kind of physical power was it that these masons were able to exert? They were not giants, after all: there are no broad-shouldered titans among the teams of workmen depicted in tomb reliefs.

We have already mentioned Colin Wilson's suggestion that the "group-mind" technique might have been instinctively applied—a combination of sheer determination of will and effort, of a belief that the gods could make the blocks lighter (encouraged, perhaps, by priests chanting), and of a concerted concentration on the maneuver in hand. We can add to this another possible factor acknowledged by everyone— the familiar phenomenon of what is known as second wind. We know that athletes involved in prolonged physical events—marathons, intensive tests of endurance, and so on—can sometimes experience a sudden transformation in their energy output and rhythm of movement, whereby they can breeze along maintaining the same momentum, but with apparently very little physical effort. Like superconductors, they suddenly reach a stage where they can "change gear," and subsequently continue to function at a much more efficient rate. In superconductors, of course, the switch from one state to another is an ultimate transition, equivalent to accelerating up to the speed of light in the time it takes to blink. Second wind appears to be a similar phenomenon, much less of a quantum leap, perhaps, than the transformation from ordinary state to superconducting state, but certainly a transcendental step in the right direction. If we suppose, therefore, that those involved in the actual construction of the Great Pyramid were able to induce in themselves, at will, something very like this, then possibly the cumulative effect of an entire, close-knit team of men working at this much more efficient rate might be sufficient in itself to "lighten their loads" considerably—even without divine intervention. Then again, second wind itself might be just one of many states of "humanness" attainable through the development of individual will. There might be a third degree, a fourth, and so on.

The final possibility, understandably too far-out to be taken seriously by most modern investigators, is that there may have been a genuine psychic factor involved—at least in some of the more difficult and demanding tasks. We are talking now of real, direct, psychokinetic effects in which the mind itself somehow resonates at frequencies powerful enough directly to influence physical quantum systems en masse, to make hard stone temporarily fluid for example.

As we have noted, yogis have consistently claimed that such things are indeed possible. Yogananda went much further, claiming that physical objects can actually be materialized at will—provided one has a will, of course—an idea that, as we have seen, has recently been touched upon by the modern researchers mentioned in previous chapters: Stanislav Grof, for example, who has tried to visualize a process whereby the mind influences the "generative matrix" of things, or David Bohm, who suggested that psychokinetic effects might be set in motion by individuals focusing on "meanings" compatible with resonances underlying the wave-functions controlling all material systems.

But, of course, even if material systems can somehow be influenced by psychokinetic energy, hollowing out the widely flared interiors of stone vases with narrow swanlike necks would still indicate a degree of craftsmanship incomprehensible to us. Let's say, for example, that we gave a modern craftsman a relatively soft material, like graphite, say, or even wood, and commissioned him or her to replicate, by any conceivable technological means, one of the stone vases examined by Hancock. Could it be done? Possibly, eventually, but such a task would inevitably involve the use of electrically powered machinery, perhaps fiber optics of the kind used in microsurgery for seeing inside the vessel, together with remote sensors to gauge the accuracy of the work in progress—and probably other custom-made tools of a kind not yet invented. If we were really to put our hypothetical craftsman to the test and ask him or her to equal the skill and dexterity of the ancient Egyptians by fashioning a vase made of actual diorite or some other extremely hard stone, we might even imagine the additional application of some form of ultrasonic cutting mechanism like that discussed by the modern toolmaker Christopher Dunn. And all this, remember, to create something that "primitive" man made by the thousands in the third millennium BCE, or even earlier.

So we have a whole array of mystifying evidence, all of it "written" in stone, that the orthodox archaeologist is unable to explain satisfactorily. However, in other evidence painstakingly "dug up" and then promptly disregarded by historians—namely the legends of the builder gods of ancient cultures—we are told time and again that "magic" was involved in the handling of stone, and that this magic somehow involved the use of music. Clearly, if there is any truth in these myths, we are considering here the use of "tools" quite unlike anything we know of or can reasonably imagine—mysterious musical devices with magical or supernatural powers. As we have noted, the only explanation currently on offer is that these techniques had a strong psychological element, perhaps something of the kind alluded to by Grof and Bohm, mental tools that Gurdjieff described as vibrations—"inner octaves"—and that his Hindu contemporary Yogananda called "creative light rays." And music and light are, of course, two of the most fundamental manifestations of the Hermetic Code, the universal symmetry first revealed by the builders themselves.

As we saw in the previous chapter, according to *The Tibetan Book of the Dead,* the "desire body" of the individual—what Rodney Collin called the soul, or the molecular body—can pass effortlessly right through rock masses, boulders, and mountains. Presumably this would also include stone vases, sarcophagi, and even great, monumental pyramids. We, of course, have no proof that the soul even exists, never mind that it might be capable of "miraculous action." But then, as Ouspensky said, phenomena of a higher order cannot be perceived in ordinary states of consciousness, so even if the world is positively teeming with these molecular shapeshifters floating effortlessly through anything that we would call "material," who in this present era of scientific rationalism would ever know?

The Tibetan Book of the Dead, however, was compiled possibly two thousand years before the modern scientific quest, before the tentative and guarded ideas of such as Copernicus or Kepler, Galileo, and Newton began seeping in through the cracks of the old, crumbling structures of the formal papal dogma. *The Tibetan Book of the Dead* is traditionally considered to be a description of the properties of the soul observed by "souls" who had personally experienced such "miracles" between reincarnations, like the Buddha himself, In any event, however we view its validity, the assertion that the "desire body" of the individual is a reality is quite straightforward and unambiguous.

We can speculate further, deeper. If these reported sightings of the soul are indeed valid and the "desire body" is endowed with what is described as "miraculous action," then what might the entity at the evolutionary stage described in *The Tibetan Book of the Dead* as the "clear-light of the ultimate reality"—Collin's "electronic" body—be capable of? According to the Tibetans, at this stage in the cycle, anything is possible.

Could it be, then, that the marvelous Egyptian artifacts found at Saqqarah were created by supernatural means, that is by people whose minds, as Collin says, were capable of "splitting the atom" and internally generating psychokinetic energies that could somehow temporarily neutralize the electromagnetic force, thus making granite or diorite soft enough to "plough through" with any moderately rigid instrument? Sadly, we may never know. We have the evidence, much of it inexplicable, that way back then something odd was in the air: exactly what, we can only speculate.

It seems reasonable to suppose, however, that the extraordinary mental and physical powers of the Egyptian elite were in some way linked to what Gurdjieff would call their level of being. As we have seen throughout the whole of this book, all roads eventually lead to the Giza necropolis, so it is to the creators of it that we must inevitably look for a final answer.

17

"Al-Chem"—
the Egyptian Way

Exactly how the Egyptians of the Old Kingdom attained such a high degree of physical and mental development is, of course, a matter of conjecture. My own view, discussed in some detail in *The Infinite Harmony,* is that they did this by following to the letter the original precepts of Thoth, a doctrine that describes a process of self-development enacted according to musical principles, resulting in the creation of a very special type of individual. However this may have been conducted in practical ways, it is clear that the system employed by the ancient Egyptians really worked, for these people succeeded in uniting harmoniously like no other nation in history. They were, as the Greeks would have termed it, homonoic in the fullest sense, people entirely of one mind, singularly dedicated to the task of transmitting their highly advanced knowledge out into the collective consciousness of the whole of mankind. And in this, as history bears witness, they succeeded admirably, for the Hermetic Code itself, the symbolic key to the Egyptian view of creation, subsequently became the blueprint from which all of the major religious doctrines of the world have been drawn.

Historians tell us that to the ancient Egyptians religion was an entire way of life, a mode of being quite unlike any code of conduct practiced today. The concept of an afterlife, of an existence in the celestial home of the gods, was very much more than just an imaginative belief system sustained by blind faith and primitive superstition. To these people, the afterlife was an attainable reality, one that could be realized by following the primary example of Osiris and Thoth and the many other deities described as the founding fathers of Egyptian culture and religion.

The symbolism invoked in the myth of the Judgment Hall of the Dead, in which Thoth/Hermes is a pivotal figure, explains the Egyptian concept of individual harmony as a passport to the afterlife very clearly.

In the collection of papyri known for convenience as *The Egyptian Book of the Dead,* it is said that the ka, or the spiritual double of the deceased, wanders through the darkness of the underworld in search of the Judgment Hall and takes on the name of Osiris in the hope of being restored to life, like Osiris himself. The subject then enters the vast Judgment Hall, where Osiris, described as having ten times the stature of the dead man's spiritual double, sits ready to oversee the proceedings. Between them is a giant pair of scales. Subsequently Anubis, the jackal-headed god (usually associated with Sirius, the "Dog Star"), and the hawk-headed Horus, son of Osiris (associated with the sun), wait to superintend the ritual. Thoth, the ibis-headed scribe (whose symbol is the moon) stands in attendance ready to record the result.

As we see, the symbolism described so far is unmistakably hermetic, an expression of the universal law of triple creation, the same law that was later encoded in toto in the later Revelation of St. John, who depicted the Woman in Heaven, the queen of creation, wearing a crown of stars (Anubis), a robe fashioned from the fabric of the sun (Horus), with her feet resting squarely on the moon (Thoth).

To continue with the underworld ritual, a single feather, symbol of the goddess Ma'at, whose name means "truth," is then placed on one pan of the scales, and the dead person's "heart" on the other. Only if the two pans remain perfectly balanced, that is, only if the individual's

"heart" is in perfect harmony with Truth/Ma'at, can the ka win the favor of Osiris and ultimately achieve immortality.

So this concept of universal harmony describes the principle of acting out harmonious sequences of conduct and development in space and time. This was the central theme of the "Egyptian way": the science of music, of alchemy, the way of the gods. As we have noted, this "way" was reflected quite clearly in the three major creation myths of the Old Kingdom, Memphite, Hermopolitan, and Heliopolitan, which all describe the miraculous appearance of an enlightened group of eight principal deities. Exactly the same underlying format, namely the octave, was also the basis of the annual performance of the sacred Osirian mysteries, the first "passion play," traditionally reenacted in the form of an eight-act drama. Indeed, it is very likely that everything the Egyptians did, whether building pyramids, enacting sacred rituals, or simply walking down a causeway, was invariably performed to the accompaniment of this universal music.

Even to this day, many orthodox Egyptologists still refuse openly to admit that the *pi* symmetry was known and used by the ancient Egyptians. In fact, we have persistently been told that they had no mathematics as such—a claim that might seem hard to reconcile with the absolute geometrical symmetry and precision of the Great Pyramid and with the exact mathematical relationships evident in the King's Chamber and the granite sarcophagus. Curiously, the dimensions of the Great Pyramid yield proportions with a value closer to "mathematical" *pi* (3.14159) than to the "classical" approximation (3.142857 rec.). However, with the original casing blocks now missing, and the whole structure shaken by a major earthquake several hundred years ago, it is impossible to determine whether the original angle of slope was intended to express the more accurate mathematical value of *pi* or its symbolic equivalent.

In any event, it was the classical convention that played the key role in Egyptian metaphysics. As we have seen, the Egyptian "model of the gods" was based on the phenomenon of light itself (after which, remember, the Great Pyramid was originally named), which modern science has since shown to be an electromagnetic manifestation of *pi*. It is an octave of resonance, with eight fundamental divisions in its overall structure: red, orange, yellow, green, blue, indigo, violet, and, of course, the transcendental white. But it also has three "primary" wavelengths: red, yellow, and blue, frequencies that make it possible to subdivide further this fundamental octave into three subsidiary scales, that is into a triple-octave format. Therefore *pi*, like light itself, is everywhere.

And so this universal symmetry—the Hermetic Code—was seen both as a model of perfection and as a description of a precise mode of being, an essentially musical system of conduct through which consciousness is, in effect, able to complete the course of its development and so transcend onto higher dimensions, greater "scales" of psychological "resonance." We have already noted a practical application of this "music in action" in the records of Old Kingdom administrative procedure, where the vizier to the pharaoh, the high priest and keeper of the mysteries, was given direct control over all twenty-two "nomes" (districts) of Upper or Southern Egypt, while his deputy, still perhaps undergoing various intermediate stages of initiation, was given subsidiary control over just seven nomes.

This unique "musical" relationship between the two priests is particularly interesting, because it brings us back to an idea discussed in earlier chapters, in which I proposed that all creative processes, whether they occur below in the microcosmic world of the self-replicating cell, or above in the double helix of the mind of the shaman or master mason, are, in fact, organic in nature. Remember, hermetic is genetic. It follows, therefore, that the process of passing on knowledge from one individual to another, from teacher to pupil, master mason to apprentice, was, in a very real sense, an organic system, one that involved disciplined, harmonious conduct and, of course, the subsequent systematic dissemination of the "immaculate" concepts by which they were guided. Being, as it were, "psychologically sound," these original concepts were quite naturally replicated faithfully, "religiously," in every succeeding generation. Thus, despite two interim periods of destructive social anarchy, the Egyptian way of life continued virtually uninterrupted for three long millennia.

This longevity, I believe, is the result of what is, in reality, an organic process, whereby the original, highly potent ideas of the gods of the First Time, exactly like successful genes in the biological heritage of dominant, evolving species, were repeatedly and faithfully "copied" in the evolving metaphysical "gene pool" of the collective Egyptian psyche.

We know that in the natural course of Darwinian evolution successful genes can survive all manner of catastrophes: ice ages, rapid meltdowns, deluges, earthquakes, cometary impacts. In the same way, the hermetic ideas we are dealing with here—the metaphysical equivalent of successful genes—have survived all kinds of social upheaval: wars, dark ages, periods of total ignorance and barbarism, inquisitions, revolutions, and so on. Therefore we are not speaking in metaphor: we are talking about organic processes of creation and evolution, both microcosmic and macrocosmic, which are identical in every way, with a difference in scale only. This, of course, is precisely what is being referred to in the hermetic dictum quoted many times before: "As above, so below." We can take this quite literally: the genetic code of the microcosm is the medium through which greater organisms evolve, and exactly the same pattern is repeated in the "cosmos" above, where the Hermetic Code describes the process by which consciousness grows and develops. There is a passage in a collection of post-Christian texts known as the Corpus Hermeticum that comes close to expressing the same idea. The god Thoth is here speaking to his son, Tat: "My son, Wisdom is the womb, conceiving in secret, and the seed is the true good."[1]

As I have said, I believe that practically everything the Old Kingdom Egyptians did was performed to the accompaniment, so to speak, of the esoteric music composed by the founding fathers of Egyptian culture, the so-called gods of the First Time. This implies, of course, that the entire Great Pyramid construction project itself was also conducted in accordance with the same principles. In other words, the whole project must have developed organically, which is to say that the Great Pyramid in effect "grew" out of the collective efforts of these very special people. We know that living organisms developing from microscopic embryos increase their bulk and complexity exponentially, two cells dividing into four, four into eight, and so on. Possibly, therefore, the building of the Great Pyramid began relatively slowly at first but, as the construction workers became more and more adept at their craft, more "in tune" with the tasks in hand and with one another, the rate at which the blocks were laid down would have increased accordingly, perhaps building up to a final crescendo of activity of a kind that we today can barely imagine. Indeed, were it not for the hard stone evidence at Giza staring us in the face—nearly two and a half million pieces of it—most people of a rational turn of mind would consider such a feat improbable, at least within the time span allowed by orthodox Egyptologists.

We have already established that the exact number of years taken to enact this remarkably harmonious performance is unknown, as indeed are the methods used, so it is not possible to explore an incidental pet theory of mine, which is that there might have been some sort of correlation between, on the one hand, the successive stages of construction and development of the structure and, on the other, the harmonic ratios of musical theory. Nevertheless, if the whole project, from start to finish, is viewed—as the Egyptians viewed almost everything—as a hermetic phenomenon, then we can say that the Great Pyramid itself, the first and foremost of the seven wonders of the ancient world, also represents the final "note" of the completed scale of enactment. And the final note of any major scale, as we know from musical theory, has transcendental properties, because it is also the first note of the greater scale above. In exactly this way, the Pyramid of Khufu/Cheops can in fact be regarded as a genuine transcendental phenomenon, whose universally harmonious proportions and alignments are, even today, five thousand years after they were created, striking strangely familiar chords in the minds of anyone prepared to take time and listen.

So we see that the Great Pyramid is in reality much more than a mere building. It is a life-bearing, organic phenomenon, an "immaculately conceived," metaphysical "gene strand" of extraordinary resilience and potency, in which is encoded the secret of life itself.

I certainly don't expect a favorable response from the orthodox Egyptology establishment regarding my musical/organic interpretation of the "Egyptian way." But this does not concern me unduly. The important point is to get one's ideas aired, to "sow the seeds," and then let nature take its course—a process in which I have a great deal of faith. If one ends up as no more than a weed in Eden, there is still the possibility of a flowering of some kind. Surely this is better than sowing nothing at all.

So, while I may not be "in sync" with orthodoxy—or even, for that matter, with the ill-defined group of "New Age" thinkers at the cutting edge of the Great Debate—everyone seems to agree on one fundamental and very important point, which is that the Egyptian civilization was unique and very special. Even orthodox historians are given to using superlatives and poetic metaphor to describe the works of the first masons of this remarkable culture.

John Romer, for example, one of the most respected authorities on ancient Egypt, describes the pyramids in a way I find particularly apt in respect of the ideas discussed in this book: "the nuclear reactors of ancient Egypt, the throne of the sun itself."[2]

In a sense, of course, there is more truth in this statement than Romer himself would care to acknowledge, for the Great Pyramid— "The Lights"—is indeed a nucleus of creative, intelligent data, an undiminishing beacon, whose illuminating beams of metaphysical "light" are, even to this day, radiating constantly out into the darker world of the ordinary human psyche.

As suggested in a previous chapter, the Giza necropolis was designed as a mirror image of the sky above the Nile Delta, and the Great Pyramid itself, as well as being the repository of the wisdom of Thoth, also functioned as a kind of ceremonial launchpad for the ascending, star-bound soul of the initiate. This vital connection with the heavenly sphere, the stellar scale of existence, is generally accepted by everyone. Romer himself expresses it: "By piling form on form the Egyptians had created a shape so dramatic that, in unison with its commanding position at the horizon, it joined heaven to earth, earth to heaven."[3]

In certain texts, the pyramids are sometimes referred to as the "Mounds of Horus"—an understandable name, given the fact that Horus himself was essentially a solar deity. There is one verse of the Pyramid Texts that describes how Horus, Osiris, and other mythical deities first initiated this whole process of transcendental evolution. "There come to you . . . the gods who are in the sky, and the gods who are on earth. They make support for you upon their arms; may you ascend to the sky and mount upon it in this its name of 'Ladder.'"[4]

The "Ladder" in question is, of course, the ladder later perceived by the Hebrew Scriptures patriarch Jacob, the "rainbow covenant" of the Israelites, the phenomenon of light.

The Egyptians, it seems, had realized long ago that light is the vehicle of consciousness, the medium through which the mind is able to transcend on to the stellar scale of existence. As I have said, they did not simply believe that this was so; they knew it, because they had firsthand experience of heaven. How else could they have possibly come to terms with such mind-boggling concepts as timelessness and infinity, concepts that, even in the earliest periods, were an integral part of Egyptian metaphysics, as the passage quoted in chapter 2 from the Old Kingdom poem referring to the god-king clearly shows: "His life-span is eternity, the borders of his powers are infinity."

It should be noted that the relatively recent ancestors of the author of this verse were supposedly primitive farmers, and that Egyptian civilization at this time was allegedly barely a couple of centuries old—younger, in fact, than our own. Yet here we have a scribe contemplating ideas of such an exalted and sophisticated nature that, were you to attempt to discuss them today with your neighbor, you might predictably be met with, at best, a glazed expression. Curiously however, in scientific circles—among quantum physicists, astrophysicists, and the like—such concepts as eternity—a timeless dimension—and the infinite, spaceless realm of the nonlocal, quantum field are common currency. Similarly, if one were able somehow to travel at the speed of light and so see the world through the "eyes" of the Holy Ghost—the photon quantum—the "heavenly" realm of the Egyptian god-king would spring magically into view. Time would be perceived to dissolve into eternity, and space would enfold into a nonlocal world of the kind observed by Ouspensky, with no borders, no "sides" to it.

Another significant feature of Egyptian metaphysics that has a distinctly modern ring to it is the idea of the constant squared being the key to all creative processes. In chapter 5, I discussed briefly the mathematical trick devised by Einstein's one-time tutor, Herman Minkowski, by which he used the value of the square of the constant (speed of light) as a means of determining the amount of pure energy stored in any given mass. As we have seen, this idea seems to have been uncannily foreshadowed by Egyptian metaphysicians, who associated "The Lights"—the Great Pyramid—with what was to become known in Ptolemaic Egypt as the Magic Square of Mercury and the number 2,080, the sum of all the numbers from 1 to 64. Sixty-four is the square of the constant number, 8, the number of full notes in the major musical scale and the number of gods involved in the early myths concerning creation. And today, of course, we find that sixty-four also is the maximum number of RNA triplet-codon combinations comprising the genetic code, the symmetry employed by DNA in the creation of all known forms of life. Furthermore, it is surely no coincidence that the Hermetic Code itself, the classical convention 22/7, can be further subdivided into three inner formulae, thus producing from the original "triple octave" a composite figure of nine octaves, sixty-four notes. As we noted also in chapter 4, this same number has even cropped up in the superstring theory of subatomic quanta, which are described rather mystifyingly as one-dimensional "strings" of vibrating energy, and which are theorized as having 64 degrees of movement associated with them.

The number 64 appears also in other ancient number systems. In the tarot for example, there are fifty-six Minor Arcana cards (the number cards) and twenty-two in the Major Arcana (the picture cards). The Major Arcana is a symbolic representation of the triple octave, an expression of the formula *pi*. And according to the law of octaves, this triple octave is also, on another scale, a single octave comprising eight fundamental notes. If we subsequently add these eight fundamental notes on to the Minor Arcana figure, we are left with the magical sixty-four. Then we have the I Ching, of course, which I discussed in the introduction of this book—an exact blueprint of the genetic code itself, with its sixty-four hexagrams and eight fundamental trigrams. Another interesting example is the old British measure of ground area—the acre—640 of which constitute a square mile.

In the last chapter I mentioned the "golden mean" proportion, denoted by the Greek letter phi, which naturally occurs in the relationship between the Great Pyramid's base and the length of its apothem or slope; that is, half the base length is in the ratio 1:1.618 with the length of the apothem. Like *pi*, phi is a naturally occurring ratio. It is expressed in a well-known series of numbers known today as the Fibonacci series, named after the thirteenth-century mathematician who first noted them. Each number in the series is the sum of the two preceding ones, like so: $1 + 1 = 2, 2 + 1 = 3, 3 + 2 = 5, 5 + 3 = 8, 8 + 5 = 13$, followed by 21, 34, 55, 89, 144, 233, and so on to infinity. If we divide any given number with the one preceding it, an approximate value for phi is obtained, which is usually rounded off to 1.618. So, for example, 233 divided by 144 gives 1.618055555556. The higher the numbers used, the greater the accuracy obtained for the value of phi.

We noted previously how this proportion has a distinctive aesthetic quality when incorporated in architecture, of which the Parthenon and the Great Pyramid itself are the best-known examples. But there is also another significant aspect of the golden mean proportion, one that has a direct bearing on the central theory of this book. It seems that phi, like *pi*, also manifests in the natural, organic world, the world created by the genetic/Hermetic Code. As examples of this, we see the developmental stages of the spiral seed-patterns of the fir cone and the sunflower. Any two of these stages taken together always correspond with two consecutive numbers of the Fibonacci series. The same is true of the spiral growth pattern of the nautilus shell. The point is that this phi relationship as it manifests in the organic world is intrinsically connected with the growth and development of spirals, helices, of which the most prominent in the whole of nature is, of course, DNA. And DNA, remember, is composed of precisely sixty-four components. We should note also that all of the other "life forms" discussed in previous chapters—principally the four-dimensional structure of the human brain and of the "solar" and the "galactic" helix—are all spirals. Possibly there are harmonic geometrical and mathematical patterns in the development and growth of all of these helical structures, but this is a question that requires more space to investigate than I can currently afford.

For me, I think the most impressive feature of this Egyptian worldview, of an infinite realm inhabited by the gods above, is the fact that these people appear to have actually devised a way for individuals to experience this alternative reality for themselves, to become "gods" in their own right. We are referring here, of course, to the way of the alchemist described by the theory of transcendental evolution, a theory based on the concept of harmonizing one's inner faculties according the principles of musical theory, and of striking metaphysical "notes" up into greater "scales" of existence.

Surely even the most skeptical observers would have to admit that the formulation of an idea as far-reaching as this, one that has practical as well as theoretical applications, is in every sense a remarkable achievement. Indeed, as I said in my last book, the Hermetic Code itself is possibly the brightest idea ever conceived by man, the original "immaculate conception." As such, this concept represents an intellectual advancement of utterly staggering proportions, one which, in terms of the kind of natural selective evolution envisaged by neo-Darwinists, can accurately be described as a genuine macromutation of the hominid mind.

To summarize: in my view the Egyptians of the early dynasties were a giant of a race, people who walked the earth with their feet firmly on the ground, but whose minds and spirits knew no physical boundaries.

They existed in the infinite cosmic ocean; they were "quantum tun-nellers," "superconductors," denizens of the plane of light above and of the quantum field—the "underworld"—below.

And their secret? How did they gain access to the nonlocal dimen-sion so effectively? How did they become conscious to such a degree that they were able to see the universe from all sides at once, from above and below, inside and out?

The myths tell us quite clearly that they did this by adopting the harmonic principles of music as a code of conduct, a systematic, "reli-gious" method of harmonious psychological development, the original tenets of which were ingeniously encoded in the "immaculate" *pi* con-vention. This, surely, is the mother and father of all disciplines. It is alchemy, the "Egyptian way," the science of the followers of the enig-matic Osiris and Thoth, civilizers of "the First Time," who taught that all creative, life-bearing processes, including the ultimate flowering of human consciousness, are products of the action of the forces described by the two fundamental laws of nature—the law of three forces and the law of octaves. The law of three, as we have seen, states that every-thing created is the result of the action of three forces: active, passive, and neutral. This is, I think, precisely what lay behind the symbolism of the three major deities of the Egyptian pantheon, the origin of the all-embracing trinity, with Osiris (male, active), Isis (female, passive), and Horus, the law-conformable (neutral) product of the union of the first two. The second fundamental law, the law of octaves, states that all things created are composed within of eightfold symmetries—hence the broader Egyptian pantheon of eight principal gods, said to have appeared simultaneously (non-locally?) on the fabled "Island of Flame."

THE FINAL ANSWER

The Egyptians are believed to have had a national motto, which in Latin translates as *memento mori*, "remember you must die." The word *die* is generally taken literally, but I suspect that there was more to it than that. After all, these people did not believe in the total extinction of the human being. They believed fervently in a life after death, a life among the stars, with Osiris, Isis, Horus, Thoth, and all the rest. So why did their national motto not reflect this belief? Why not "remember you can live forever?" One can only assume that these people did not need reminding of what to them was the self-evident reality of the afterlife. Old Kingdom Egyptians were almost totally preoccupied with it, as the myths and the precise, star-bound alignments of their architecture clearly show. The reference to "dying," therefore, may have some other, more esoteric meaning, and I suspect that this was precisely the same meaning as that alluded to in the passages from Gurdjieff's book of aphorisms mentioned in the last chapter, one of which read, "When a man awakes he can die; when he dies he can be born." *Memento mori,* therefore, was probably intended to remind initiates not of their mortality, but of the way in which immortality can be achieved; that is, by dying to the illusory, material world, by regularly adopting a passive role in the cosmic scheme of things. There is a well-known biblical quotation that expresses the very same principle: "Except a corn of wheat fall into the ground and die, it abideth alone; but if it die, it bringeth forth much fruit." The organic inference here is particularly appropriate, because genetic processes, as we have seen, are hermetic; therefore "fruits" of any kind, whether above in the mind or below in the living cell, are created in exactly the same alchemical way.

So "dying" in life (meditating, making oneself receptive to greater cosmic influences) was seen as a way of preparing individuals for death as we think of it, a natural event, which to the Egyptians was seen not as a terminal event but rather as an organic transition in an ongoing evolutionary process. We might call this transition a macromutation of the human spirit, an ultimate, mind-altering metamorphosis, through which consciousness transcends on to an infinitely greater scale of existence. This is the scale alluded to in the symbolism of the two-winged caduceus, the magic wand of Hermes, a graphic representation of the greater "double helix" in the sky. This principle is clearly expressed in this verse from the text known as the Corpus Hermeticum:

> Do you not know, Asclepius, that Egypt is made in the image of Heaven, or so to speak more exactly, in Egypt all the operations of the powers which rule and work in Heaven have been brought down to earth below? Nay, it should be said that the whole Cosmos dwells in this land and its temples.[6]

So the Giza necropolis was designed as a kind of mirror image of the Egyptian Duat, of the sky, principally to emphasize humankind's star-bound destiny. It is significant that the word *Duat* also meant "underworld." Now, perhaps, we can understand why. Above and below—the plane of light and the quantum field—are one and the same nonlocal dimension. And incredible as it may seem, the Egyptians appear to have been aware of this.

Continuing for the moment with our organic perspective, it is evident that these people somehow succeeded in breaking free from the Darwinian mode of evolution common to all, and quite literally macromutated, evolved transcendentally, into a nation united, into a greater, single, homonoic "organism." What we are trying to envisage here is a kind of metaphysical "chromosome," a living, multidimensional structure, whose life-bearing data—ideas, precepts, concepts, rituals, and myths—were designed or created solely to build, on a macrocosmic scale, even greater organisms, "gods" if you will, "Tetrads in the sky."

We, today, are the inheritors of these metaphysical "genes" and, although our general mode of evolution is characteristically Darwinian—"naturally selective"—I believe that buried within the collective consciousness of the human race there remains an underlying tendency to evolve transcendentally, just as the Egyptians did. As we have seen, these enigmatic people not only evolved into a race apart, they left behind them all the data required for us to follow in their wake. They planted "seeds" as they passed through this world, seeds of wisdom, of symbol, myth, and legend; seminal ideas, which, over the millennia, have periodically germinated and come to fruition, and which today are once again beginning to produce a whole "new" variety of conceptual flora.

Modern science, for example, which seems to me to have been born out of an instinctive need for the human mind to overcome the desolate, stultifying climate of the Inquisition, is now poised to enter its transcendental phase. Accordingly our attention is once again turning to things "above," to the cosmos itself, and to things "below," to the quantum field and the nonlocal realm being explored in scientific communities worldwide.

The early pioneers of the modern scientific movement—Newton, Galileo, Copernicus, Kepler, and so on—began this present phase of metaphysical growth when they started to observe the heavenly bodies and to understand the forces controlling them. The ensuing process of scientific enquiry culminated in the ideas of Albert Einstein, whose own attention was eventually to focus, perhaps inevitably, on the constant light of the sun. In a sense, therefore, through the concepts of this modern genius, the great Egyptian sun-god Ra has triumphantly returned, bringing with him a glimmer of understanding, a timely recognition of the eternal, spaceless dimension in which he reigns supreme.

So the ancients' description of the constant realm of the god-king, formulated by people to whom, one suspects, "transpersonal experiences" were readily accessible, was subsequently reborn under its modern guise of Special Relativity, the theory that finally turned logical thought upside down, and that ultimately gave rise to the "new" scientific vision of a nonlocal universe.

But, as we have noted, this modern "gene-strand" of ideas is actually a mutated form of the original "immaculate conception." In reality the basic components of the Egyptian way, exactly like the dancing genes in the DNA of a newly fertilized ovum, have simply been "jiggled about," but they remain essentially the same components, the same genes. Even in King Solomon's day, it was understood that the esoteric traditions of the Judaic religion were simply echoes of a much older theme: "and there is no new thing under the sun. Is there anything whereof it may be said, See, this is new? It hath been already of old time, which was before us."[7]

Throughout recorded history there have been other, quite distinct, mutations in human thought, characteristic variations in the evolving species of nations. The Chinese, the Indian, Persian, Greek, Judaic, Christian, and Islamic codes of life—all of these metaphysical "creatures" have been born and have thrived in their own day as dominant gene-strands. Today these same genes are in a passive or recessive mode; scientific ideas and concepts have now superseded them. It may be argued that the Christian and Islamic traditions are still dominant, active, but I would suggest that this is due largely to extreme fundamentalist elements of a type that Jesus and Muhammad, both of whom were relative paragons of compassion and tolerance, would be unlikely to countenance if they were around today. But, in any event, all of the major religions and esoteric traditions are still there, still alive (literally) in the great gene pool of human consciousness. In subsequent generations they may even become dominant again, may each undergo a sudden resurgence or renaissance, as the human brain continues to develop and to adapt to environmental variables.

The obvious conclusion to be drawn from this ongoing "meta-biological" process of thought is that the ancients had it exactly right, that the entire cosmos—the real universe, as opposed to the four-dimensional physical shadow perceived in our ordinary states of awareness—is indeed a living, breathing creature like you, whose life-blood is none other than consciousness itself. This in turn suggests that the whole universe, like any organic body, is pulsating throughout with life.

Sri Aurobindo said that if a single point in the universe were unconscious, then the whole universe would have to be unconscious. Scientifically we can interpret this to mean that if the "mindlike" qualities of the photon or the electron were removed, if "nonlocal quantum correlations" were to cease, the whole cosmos would become a dark and lifeless void. Fortunately the great ancient sun god is currently alive and well and gloriously omnipotent, and as long as this universal archetype continues to inhabit our dreams and to be the principal vehicle of our perceptions, the human race, it seems, will never be alone.

So it is very likely that science fiction has been nearer to fact than many people imagine and that there are "aliens" out there. If the universe is a zoon, an immense, six-dimensional creature going around by the name of God, there must be. But these extraterrestrials, no matter what form they might take, are our brothers and sisters, metaphysical "protein-builders" just like ourselves, created by, and acting under, the direct influence of "gods" of star-strung, serpentine "chromosomes."

Now, here's a thought. If we are ultimately to turn science fiction into reality and communicate directly and coherently with our extraterrestrial counterparts across billions of light years of space, then the connection, one suspects, will somehow have to be made, not through the use of impossible-to-build "warp factor" starships, or hypothetical "wormholes" in the curved fabric of space-time, or even radio waves, but through the metaphysical frequencies of the nonlocal, subquantum ("underworld") channel of communication. The Egyptians, of course, have already made contact with other beings; they have "died" and journeyed to the underworld and passed the ultimate test of truth. And so, too, have all the other remarkable teachers of hermetic wisdom mentioned in this book, individuals whose thoughts, ideals, and concepts still flourish unceasingly in the collective consciousness of the human race as it grows, a shimmering, multidimensional pyramid of resonant data, up toward the heavens. These great souls have already been born into spirit; they are, in a sense, already "out there," communing with the godlike inhabitants of the starry world, waiting patiently for us to join them in the celestial celebration that never ends, a party to which, it seems, we have all been cordially invited.

So when you think you're ready, you might care to rendezvous at the Giza terminal. Even if you get there only in your wilder dreams, it all adds up. The more positive thought patterns we transmit out into the nonlocal energy field (the plane of light, the "book of life"), the more we will ultimately get out of it. Our input, however, if it is to have any lasting effect, will have to be homonoic, that is, conducted through a genuine union of minds. Like the pyramid builders we will all have to pull together and start integrating in a true spirit of cooperation and openmindedness. Presumably the cumbersome ego will have to be completely discarded. Remember the feather on the balance in the Judgment Hall of the Dead, the symbol of Truth. What earthbound ego could possibly pass such a test of its real substance? None.

So think of the stages of evolution enacted in the metamorphosis from caterpillar to butterfly as the evolving entity sheds its dense, gravity-bound chrysalis and ultimately flies up into the sky, to a new life. Perhaps, through living simply in compliance with the basic laws and forces of nature, this could be you, the eagle-beast of Revelation, soaring to places ordinary mortals can only dream about.

Collectively, as the human race fast approaches the new precessional Age of Aquarius, we are facing a crucial and momentous decision: either we evolve in harmony, transcendentally, united as one, in a higher dimension, a greater scale of being, or we remain fragmented, divided, isolated in time and space, a timid, provincial race dead from the neck up, enslaved by economic obligations, eking out a meager existence on a sad little planet littered with fossil dinosaurs, dodos, and countless other extinct species.

For my money, and for the sake of all around me, I feel strongly inclined to go for the former option, to follow in the footsteps of the Egyptian high priest. We can do it if we want to. It is basically a state of mind, but one that, as the concept of the eternal trinity implies, can only manifest through the harmonious interaction of the three fundamental forces of nature: active, passive, and neutral, and in that order. Ordinary thought processes switch from active to neutral and back again, endlessly. In this lies our greatest folly, because the genuine passive element is always absent, which means that the mind is never fully receptive, never able to assimilate external data in sufficient quantities to stimulate growth. Remember the pyramid ritual, the opening of the mummy's mouth at the foot of the southern shaft of the Queen's Chamber, aligned to Sirius, star of Isis, the passive force of the trinity. This is alchemy pure and simple, a description of the vital process of opening the mind, of "waking it up," so to speak. This, of course, is precisely what genuine and sincere prayer, meditation, and numerous other yogic practices were designed to do—to introduce the passive element into the processes of mind, without which there can be no rhythm, no real harmony. So here's a tip: keep your "sabbath," your period of "rest"—you can't be fully in tune with nature without it.

Significantly, we need only look to the microcosm, to the evolution of DNA, to realize that the Egyptians themselves must have "sung" like proverbial angels, for the "pyramid ritual" is, in fact, performed repeatedly by all chromosomes, the "minds" of the biomolecular world. When the chromosome is ready to act, it first relaxes the tension of one of its two nucleotide chains; that is, it becomes temporarily passive. This, effectively, opens up the double-helix structure, causing the paired bases within it to separate, at which point, something quite "magical" occurs. Free nucleotide bases floating around in the surrounding cytoplasmic membrane are taken in by the chromosomes. The chromosome then combines these bases into "triple-octave" units—RNA codons, the microcosmic equivalent of concepts—and subsequently ejects them again to carry out a specific evolutionary function, which is to act as templates for the manufacture of amino acids, the building blocks of life.

And what do all self-respecting, self-replicating cells do with these building blocks? They build "pyramids," of course: living ones, immense, six-dimensional organic structures capable of building even greater pyramids . . .

Notes

INTRODUCTION

1. P. D. Ouspensky, *In Search of the Miraculous,* 124.
2. It might be argued that seeing the numbers 7 and 8 as interchangeable makes the identification of patterns too easy, but I think that any "natural" configuration that conforms to this universal symmetry, that coincides at the key "points of entry," as it were (for example, points 3, 4, 7, 8, 22, 64, and combinations thereof)—musical, geometric, genetic, conceptual—is, I think, valid. So the 838 symbolism of the I Ching, or the "chessboard" ground plan of the Giza Necropolis, is essentially expressing the same principle as the 93711 format of the triple-octave "squared," because the product of each is 64. Similarly the number 64 in the genetic code is obtained, not through an 838 format, or 93711, but through 43434—again, with a product of 64. So these apparently disparate patterns do map one onto another, but only at certain crucial points. One would not expect exact superimpositions to be visible at every level, because the universe is continually evolving, constantly in flux. But as long as the various symmetries link in at these main "points of entry" the Hermetic Code is valid. If anything, the fact that the code can be directly linked to all of these various symmetries—and many others found throughout the natural world—is compelling evidence of its extraordinary dynamism and universality. This is precisely what one would expect of a "theory of everything."
3. Michael Hayes, *The Infinite Harmony,* 17.

CHAPTER 1. THE SACRED CONSTANT: THE "JEWEL IN THE CROWN"

1. William R. Fix, *Pyramid Odyssey,* 108.
2. R. A. Schwaller de Lubicz, *Sacred Science,* 86.
3. Colin Wilson, *From Atlantis to the Sphinx,* 78.
4. Edouard Naville, "Excavations at Abydos," cf. Corliss, 325.
5. Andrew Collins, *Gods of Eden,* 11.
6. Rand Flem-Ath and Colin Wilson, *The Atlantis Blueprint,* chapter 3, "The Giza Prime Meridian."
7. Peter Tompkins, *Secrets of the Great Pyramid,* 287–382.
8. Stan Gooch, *Cities of Dreams,* 99–100.
9. Ibid., chapter 10.

CHAPTER 2. A DIFFERENT WAY OF SEEING

1. Colin Wilson, *From Atlantis to the Sphinx,* 9.
2. Colin Wilson, *The War Against Sleep,* 89.
3. Colin Wilson, *From Atlantis to the Sphinx,* 10.
4. A. Erman, *Pyramid Texts, the Literature of the Ancient Egyptians,* 4f.
5. Colin Wilson, *From Atlantis to the Sphinx,* 242.
6. Ibid., 242.
7. R. A. Schwaller de Lubicz, *Sacred Science,* chapter entitled "Magic, Sorcery, Medicine."
8. Colin Wilson, *From Atlantis to the Sphinx,* 246.

CHAPTER 3. MUSIC OVER MATTER

1. Carl G. Jung, *Memories, Dreams, Reflections,* 178–79.
2. John Bierhorst, *The Mythology of Mexico and Central America,* 8.
3. Harold Osbourne, *Indians of the Andes: Aymaras and Quechuas,* 64.
4. Mark Henderson, *The Times,* February 16, 2004.
5. Paramhansa Yogananda, *Autobiography of a Yogi,* 316.
6. Andrew Collins, *Gods of Eden,* 66–70.
7. Graham Hancock, *Fingerprints of the Gods,* 262.
8. Andrew Collins, *Gods of Eden,* 82.

9. Ibid., 77–78.

10. Peter Tompkins, *Secrets of the Great Pyramid,* 101–3.

CHAPTER 4. THE ELECTRON AND THE HOLY GHOST

1. Pierre Speziali (ed.), Einstein–Besso Correspondence, 1903–1955, p. 538.

2. Michael Talbot, *The Holographic Universe,* 51.

3. P. D. Ouspensky, *In Search of the Miraculous,* 176.

4. Paul Davies, *Other Worlds,* 68.

5. Timothy Ferris, *The Whole Shebang,* 224.

6. Colin Wilson, *The Strange Life of P. D. Ouspensky,* 54.

7. Ibid., 50.

8. Satprem Satprem, *Sri Aurobindo, or the Adventure of Consciousness,* 219.

9. Paramhansa Yogananda, *Autobiography of a Yogi,* 316.

CHAPTER 5. FURTHER LIGHT

1. Graham Hancock, *Fingerprints of the Gods,* 178.

2. Ralph Ellis, *Thoth, Architect of the Universe,* 3.11.

CHAPTER 6. LIVE MUSIC

1. Michael Hayes, *The Infinite Harmony,* 221–30.

2. Richard Dawkins, *River Out of Eden,* xi.

3. Ibid., 52.

4. Ibid., 33.

5. Richard Dawkins, *The Selfish Gene,* 192.

6. Ibid., 199.

7. Ibid., 197.

8. Michael Hayes, *The Infinite Harmony,* 154–57.

9. Richard Dawkins, *The Selfish Gene,* 196.

10. Graham Hancock, *Fingerprints of the Gods,* 95.

11. James Shrieve, *The Neanderthal Enigma,* 69.

12. Stan Gooch, *Cities of Dreams,* 49–53.

13. Michael Hayes, *The Infinite Harmony,* 27–39.

14. Linda Jean Shepherd, *Lifting the Veil, The Feminine Face of Science,* 215.
15. Richard Dawkins, *River Out of Eden,* 142.

CHAPTER 7.
EXTRATERRESTRIAL DNA

1. Rodney Collin, *The Theory of Celestial Influence,* 342.
2. Michael Hayes, *The Infinite Harmony,* 222.
3. Rodney Collin, *The Theory of Celestial Influence,* 81.

CHAPTER 8.
INTERSTELLAR GENES AND
THE GALACTIC DOUBLE HELIX

1. Robert Temple, *The Sirius Mystery,* 3.
2. Ibid., 55.
3. Ibid., 24.
4. Ibid., 25.
5. Michael Hayes, *The Infinite Harmony,* 81–93.
6. Robert Temple, *The Syrius Mystery,* 28.
7. Ibid., 29.

CHAPTER 9. THE HERMETIC UNIVERSE
OF ANCIENT TIMES

1. Genesis 1:1.
2. Ibid., 1:2.
3. Ibid., 1:3–4.
4. F. Max-Muller, *The Laws of Manu,* 1:8–9.
5. George Smoot, *Wrinkles in Time,* 272.

CHAPTER 11.
THE FATE OF THE UNIVERSE

1. Paul Davies, *The Last Three Minutes,* 67–68.

CHAPTER 12. INNER OCTAVES

1. P. D. Ouspensky, *In Search of the Miraculous*, 86.
2. Ibid., 81.

CHAPTER 13.
THE HOLOGRAPHIC PRINCIPLE

1. John Blofeld, *Tantric Mysticism of Tibet*, 61–62.
2. P. D. Ouspensky, *In Search of the Miraculous*, 88.
3. Genesis 9:6.
4. *Compton's Interactive Encyclopaedia*, "Orpheus" entry.

CHAPTER 14. QUANTUM PSYCHOLOGY:
THE "NONLOCAL" BRAIN

1. Michael Talbot, *The Holographic Universe*, 122.
2. P. D. Ouspensky, *In Search of the Miraculous*, 88.
3. Ibid., 262.
4. Ibid., 265.
5. Ibid., 265–66.
6. Stanislav Grof, *Beyond the Brain*, 91.
7. Colin Wilson, *The Strange Life of P. D. Ouspensky*, 48.
8. Ibid., 48.
9. Paramhansa Yogananda, *Autobiography of a Yogi*, 318.
10. Ibid., 319.

CHAPTER 15.
QP2: THE UNIVERSAL PARADIGM

1. Revelations 12:1.
2. P. D. Ouspensky, *In Search of the Miraculous*, 217.
3. Ibid.
4. Paramhansa Yogananda, *Autobiography of a Yogi*, 315.
5. Ibid.
6. W. Evans-Wentz, *The Tibetan Book of the Dead*, 158–59.

7. Ibid.

8. Ibid.

9. Ibid., 95–96.

10. Rodney Collin, *The Theory of Eternal Life*, 37.

CHAPTER 16. THE SHAPESHIFTERS

1. Adrian Recinos, *Popul Vuh, The Sacred Book of the Ancient Quiche Maya*, 168–69.

2. Ibid., 169.

3. Ibid.

4. Graham Hancock, *Fingerprints of the Gods*, 337.

5. Ibid., 333.

6. Mark Lehner, *Secrets of the Lost Empires*, 93.

CHAPTER 17. "AL-CHEM"—THE EGYPTIAN WAY

1. Anon., *Material for Thought*, no. 7, 1.

2. John Romer, *Romer's Egypt*, 28.

3. Ibid., 65.

4. R. O. Faulkner (trans.), *Ancient Pyramid Texts*, 227.

5. David Furlong, *Keys to the Temple*, 79.

6. Corpus Hermeticum, Asclepius III 246 (see under Copenhaver in bibliography).

7. Ecclesiastes 1:9–10.

Bibliography

Anon. *Material for Thought, Excerpts from the Hermetica*. San Francisco: Far West Press, 1977.

Bauval, R., and A. Gilbert. *The Orion Mystery*. London: Heinemann, 1994.

Bible, The. King James version.

Bierhorst, John. *The Mythology of Mexico and Central America*. New York: Morrow, 1990.

Blofeld, John. *Tantric Mysticism of Tibet*. New York: E. P. Dutton, 1970.

Collin, Rodney. *The Theory of Celestial Influence*. London: Robinson & Watkins, 1980.

———. *The Theory of Eternal Life*. Boston: Shambhala, 1984.

Collins, Andrew. *Gods of Eden*. London: Headline, 1998.

Copenhaver, Brian P. *Corpus Hermeticum*. Cambridge: Cambridge University Press, 1992.

Davies, Paul. *The Last Three Minutes*. London: Weidenfeld & Nicolson, 1994.

———. *Other Worlds*. London: Penguin, 1990.

Dawkins, Richard. *The Blind Watchmaker: Why the Evidence of Evolution Reveals a Universe Without Design*. London: Penguin, 1988.

———. *River Out of Eden*. London: Orion, 1995.

———. *The Selfish Gene*. Oxford: Oxford University Press, 1989.

De Santillana, Giorgio, and Hertha von Dechend. *Hamlet's Mill: An Essay Investigating the Origins of Human Knowledge and Its Transmission through Myth*. Boston: David R. Godine Publisher, 1992.

Dunn, Christopher P. *The Giza Power Plant: Technologies of Ancient Egypt*. Rochester, Vt.: Bear & Co., 1998.

Ellis, Ralph, *Thoth, Architect of the Universe*. Dorset: Edfu Books, 1997.

Erman, A. *Pyramid Texts, the Literature of the Ancient Egyptians*. London: A. M. Blackman, 1927.

Evans-Wentz, W. *The Tibetan Book of the Dead*. Oxford: Oxford University Press, 1974.

Faulkner, R. O., trans. *Ancient Pyramid Texts*. Oxford: Oxford University Press, 1969.

Ferris, Timothy. *The Whole Shebang*. London: Weidenfeld & Nicolson, 1997.

Fix, W. R. *Pyramid Odyssey*. Toronto: J. James, 1978.

Flem-Ath, Rand, and Colin Wilson. *The Atlantis Blueprint*. London: Little Brown, 2000.

Gooch, Stan. *Cities of Dreams*. London: Rider, 1989.

Graves, Robert. *The White Goddess: A Historical Grammar of Poetic Myth*. London: Faber & Faber, 1999.

Grof, Stanislav. *Beyond the Brain*. Albany: State University of New York Press, 1985.

Gurdjieff, G. I. *Beelzebub's Tales to His Grandson*. London: E. P. Dutton, 1964.

Hall, Edward T. *The Dance of Life: The Other Dimension of Time*. New York: Anchor, 1984.

Hancock, Graham. *Fingerprints of the Gods*. London: Heinemann, 1995.

Hapgood, Charles H. *Maps of the Ancient Sea Kings: Evidence of Advanced Civilization in the Ice Age*. London: Souvenir Press, 2001.

Hayes, Michael. *The Infinite Harmony*. London: Weidenfeld & Nicolson, 1994.

Jung, Carl G. *Memories, Dreams, Reflections*. London: Fontana, 1995.

Lehner, Mark. *Secrets of the Lost Empires*. London: BBC, 1996.

Max-Muller, F. *The Laws of Manu*. Delhi: Motilal Banarsidas, 1964.

Moskvitin, Jurij. *Essay on the Origin of Modern Thought*. Athens: Ohio University Press, 1974.

Narby, Jeremy. *The Cosmic Serpent*. London: Weidenfeld & Nicolson, 1998.

Naville, Edouard. "Excavations at Abydos." Smithsonian Institution Annual Report, 1914.

Osbourne, Harold. *Indians of the Andes: Aymaras and Quechuas*. London: Routledge and Kegan Paul, 1952.

Ouspensky, P. D. *A New Model of the Universe.* Mineola, N.Y.: Dover Publications, 1997.

———. *In Search of the Miraculous.* London: Routledge & Kegan Paul, 1976.

Paramhansa Yogananda. *Autobiography of a Yogi.* Los Angeles: Self-Realization Fellowship, 1977.

Recinos, Adrian. *Popul Vuh, the Sacred Book of the Ancient Quiche Maya.* Athens: University of Oklahoma Press, 1991.

Romer, John. *Romer's Egypt.* London: Michael Joseph, 1982.

Satprem Satprem. *Sri Aurobindo, or the Adventure of Consciousness.* New York: Institute for Evolutionary Research, 1984.

Schwaller de Lubicz, R. A. *Sacred Science.* Rochester, Vt.: Inner Traditions, 1988.

Shepherd, Linda Jean. *Lifting the Veil: The Feminine Face of Science.* Boston: Shambhala, 1993.

Shrieve, James. *The Neanderthal Enigma.* London: Viking, 1996.

Smoot, George. *Wrinkles in Time.* New York: William Morrow, 1994.

Speziali, Pierre, ed. Einstein–Besso Correspondence, 1903–1955. Paris: Hermann, 1972.

Talbot, Michael. *The Holographic Universe.* London: Grafton Books, 1991.

Temple, Robert. *The Sirius Mystery.* London: Century, 1998.

Tompkins, Peter. *Secrets of the Great Pyramid.* London: Allen Lane, 1971.

VandenBroeck, André. *Al-Kemi: Hermetic, Occult, Political and Private Aspects of R. A. Schwaller de Lubicz.* Herndon, Va.: Lindisfarne Press, 1990.

West, John Anthony. *Serpent in the Sky: High Wisdom of Ancient Egypt.* Wheaton, Ill.: Quest Books, 1993.

Wilson, Colin. *From Atlantis to the Sphinx.* London: Virgin, 1996.

———. *The Strange Life of P. D. Ouspensky.* London: Aquarian Press, 1993.

———. *The War against Sleep.* London: Aquarian Press, 1980.

Spalding

Index

Annanuki ?